Engineering Chemistry

I

Engineering Chemistry

I

R Jayaprakash MSc, MPhil (Phd)

Lecturer
Department of Chemistry
Dhaanish Ahmed College of Engineering
Padapai 601301
Tamil Nadu

CBS Publishers & Distributors Pvt Ltd

New Delhi • Bengaluru • Pune • Kochi • Chennai

Engineering Chemistry I

ISBN: 978-81-239-1938-6

First Edition: 2011

Published by Satish Kumar Jain and produced by Vinod K. Jain for
CBS Publishers & Distributors Pvt Ltd
4819/XI Prahlad Street, 24 Ansari Road, Daryaganj,
New Delhi 110 002, India.
Ph: 23289259, 23266861/67 Fax: +91-011-23243014
Website: www.cbspd.com
e-mail: delhi@cbspd.com; cbspubs@vsnl.com; cbspubs@airtelmail.in.

Branches

- Bengaluru: Seema House 2975, 17th Cross, K.R. Road, Banasankari 2nd Stage, Bengaluru 560 070, Karnataka
  Ph: +91-80-26771678/79 Fax: +91-80-26771680 e-mail: bangalore@cbspd.com
- Pune: Bhuruk Prestige, Sr. No. 52/12/2+1+3/2 Narhe, Haveli (Near Katraj-Dehu Road Bypass), Pune 411 051, Maharashtra
  Ph: 020-64704058. 64704059 e-mail: pune@cbspd.com
- Kochi: 36/14 Kalluvilakam, Lissie Hospital Road, Kochi 682 018, Kerala
  Ph: +91-484-4059061-65 Fax: +91-484-4059065 e-mail: cochin@cbspd.com
- Chennai: 20, West Park Road, Shenoy Nagar, Chennai 600 030, Tamil Nadu
  Ph: +91-44-26260666, 26208620 Fax: +91-44-45530020 email: chennai@cbspd.com

Printed at India Binding House, Noida, UP

Preface

This book completely covers the chemistry curricula prescribed for the first semester engineering degree courses of Anna University, Chennai, Trichy, Coimbatore and Trinalveli, affiliated engineering colleges and all engineering departments.

The book *Engineering Chemistry I* has been written with applications of chemistry towards technology. It depicts points with good diagrammatic representations for all the topics so that the readers can easily understand the subject.

This book is 100% result oriented and it explains the detailed questions by exam presentation method.

I hope that the book will be useful for all the students as well as the teachers. Finally, I request students and faculty to kindly send me their comments and suggestions for the improvement of this book and for the student's development.

R Jayaprakash
e-mail ID: krjp_1977@yahoo.co.in
Ph : 94448 95313, 04183 244169

Acknowledgements

My sincere thanks to well-wishers Dr I Mohammad Bilal, Professor and Head, my guide Dr D Easwaramoothy, Professor, and Dr S Kutti Rani, Professor, Department of Chemistry, BS Abdur Rahman University, Vandalur. My heartiest thanks to my co-guides, Dr M Palanichamy, Professor, and Dr N Rajendran, Professor, Department of Chemistry, Anna University, Chennai. My sincere thanks to Dr SN Narasimman, Director, Asthagiri Herbal Research Foundation, Chennai.

My sincere thanks to my beloved Alhaj K Moosa, Chairman and Founder of Dhaanish Ahmed College of Engineering, Mr Mohammad Hafeez, Vice-chairman and Mr Kadar Shah, enthusiastic Secretary, who have encouraged me to write the book. I am grateful to Dr S Poornachandra, Principal, Dhaanish Ahmed College of Engineering, who inspired me to write this book.

My warm thanks to Dr K Ravichandran, Department of Rubber Technology, MIT Campus, Chennai.

My sincere thanks to my colleagues and friends who have given me valuable suggestions during the development of the material. My grateful thanks to Mr R Mohan Raj (JRF) who helped me in designing and typing this material, BS Abdur Rahman University, Vandalur, Chennai.

R Jayaprakash

Contents

1

Water Technology

OBJECTIVES

Water is used for domestic purposes. Nowadays preservation of water sources and characters maintenance is an important one. We are using water for different purposes, e.g. Drinking, Industrial uses. Therefore, we have to know about the properties, purification methods. Technologies are related to water purification and preservation.

1.1 INTRODUCTION

75% of earth is surrounded by water. Our major source is the ocean. Nevertheless, the nature of water is salty. We cannot utilize it for domestic purposes. Our population is more but water availability is less. To fulfill our requirements seawater should be purified. Our important water requirement areas are:

- Agricultural
- Industrial
- Domestic or Waste water

The purest form of water is rain. While falling down it mingles with different gases and impurities. So, we have to identify the water properties:

Molecular formula of water	H_2O
Molecular weight	18.0153 g/mole
Density	$0.998 g/cm^3$
Melting point	0°C, 273 K
Boiling point	100°C

Water parameters are important for commercial and domestic uses

For an example: Ice cube formation. Time taken for pure water to form an ice cube is greater than that of impure water. The same manner is for melting also. Pure ice cube melting time is greater than that of impure ice cube.

Flow chart 1.1

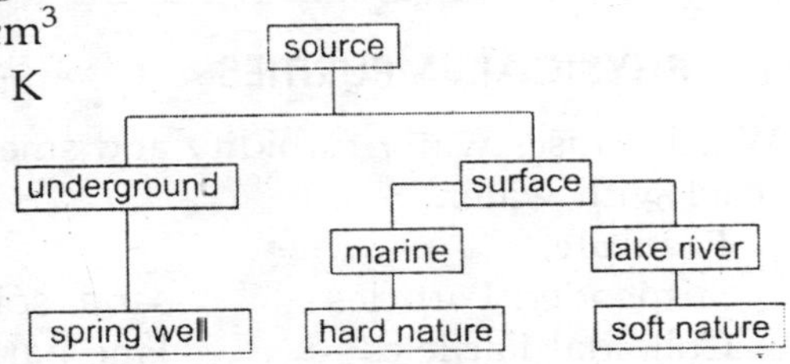

Sources of Water

Based on the nature of availability, water can be classified as depicted in Flow chart 1.1.

Impurities

The particles which pollute water are called water impurities. Based on their nature impurities can be of three types as depicted in Flow chart 1.2.

Flow chart 1.2

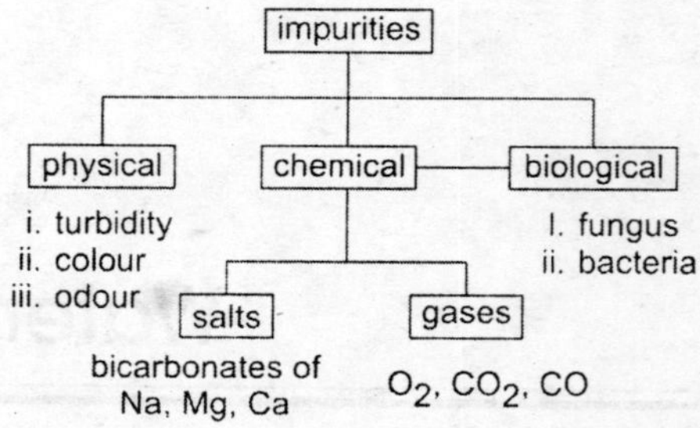

Water Treatment

For domestic usage purpose, raw water is treated with some chemicals and purified, this process is called water treatment.

Water Technology

Using Engineering and Technology Methods, impure raw water converted to Domestic usage water is called water Technology.

1.2 CHARACTERISTICS OF WATER

Based on the impurities chemical, physical properties named as follows:

Chemical characters	***Physical characters***
Alkalinity	Colour
Hardness	Turbidity
Softness	Taste and smell
Acidity	Electrical conductivity
Ph	
Chloride	
Fluorides	
Sulphates	
Nitrogen compounds	
Dissolved oxygen	
Biological oxygen demand	
Chemical oxygen demand	

1.3 PHYSICAL IMPURITIES

Which causes water turbidity and smell is called physical impurities. These do not react with water.

Example,

Suspended Particles — sand, soil, vegetable and animal matters

Colloidal Particles — Fine particles
Organic waste products
These cause colour, odour and taste

1.4 CHEMICAL IMPURITIES

Chemicals which change the nature of water and react with water are called chemical impurities. Based on the state of the chemical, it is classified into two types:

1.5 DISSOLVED SALTS (SOLID)

- Salts of Na, K, Ca and Mg
- Bicarbonates, Sulphates and Chlorides

  Example, $NaHCO_3$, Na_2CO_3, Na_2SO_4, $MgSO_4$, $CaCl_2$ and $CaSO_4$

1.6 DISSOLVED GASES

These gases react with water. It causes acidity to the water. Gases react with H^+ ion and causing acidity, it causes corrosion on metal surfaces.

Example, O_2, CO_2, H_2S, Cl_2 and NO_2

$$H_2O \longrightarrow H^+ + OH^-$$

$$H^+ + Cl_2(gas) \longrightarrow HCl\ (Acid)$$

$$H_2O + NO_2 \longrightarrow HNO_3\ (Acid)$$

1.7 BIOLOGICAL IMPURITIES

These are mingled with water, the non-visible microorganisms are called biological impurities. These cause harmful diseases.

Example, Fungus, Bacteria and Virus

Diseases: Malaria, Typhoid and Dysentery.

Biological Oxygen Demand

This is because of biological impurities. Oxygen demand because of biological impurities is called Biological Oxygen Demand.

Chemical Oxygen Demand

The oxygen demand because of chemical impurities is called Chemical Oxygen Demand.

In waste water — COD > BOD

1.8 ALKALINITY

Definition

It is the tendency of water to accept H^+ ions. Alkalinity is a measure of the ability of water to neutralize the acid.

$$\underset{\text{Acid}}{H^+} + \underset{\text{Base}}{OH^-} \longrightarrow \underset{\text{Neutral}}{H_2O}$$

HCO_3^-, OH^- and CO_3^{2-} ions are reason for alkalinity

Reason for Alkalinity

Alkalinity of water due to the formation of Carbonates, Bicarbonates and Hydroxides in water. First, metal Mg or Ca forming its Carbonates:

$$MgO + CO_2 \longrightarrow MgCO_3 \xrightarrow{H_2O + CO_2} Mg(HCO_3)_2$$

$$CaO + CO_2 \longrightarrow CaCO_3 \xrightarrow{H_2O + CO_2} Ca(HCO_3)_2$$

Natural water containing more alkalinity. These Mg and Ca reacting with CO_2 and forming alkalinity (Calcium and magnesium Carbonate).

Why we have to calculate alkalinity? or Reasons for alkalinity measurement.

1. To control the rate of corrosion. Water containing basic nature reacts with metal and it degrade the metal surface
2. To calculate the amounts of lime (Calcium hydroxide) and Soda (Sodium carbonate) required for the removal of hardness
3. For conditioning inside the boiler feed water

1.9 TYPES OF ALKALINITY

Based on the alkalinity causing Anions (or) Negative ions, alkalinity is classified into 3 types:

1. Alkalinity due to OH^- ions is called Hydroxide alkalinity (Strong alkalinity)
2. Alkalinity due to CO_3^{2-} ions is called Carbonate alkalinity (Medium alkalinity)
3. Alkalinity due to HCO_3^- ions is called Bicarbonate alkalinity (Weak alkalinity)

Temporary Alkalinity

The alkalinity due to (HCO_3^-) bicarbonate ions is called weak or temporary alkalinity.

Caustic Alkalinity

If the alkalinity due to the presence of CO_3^{2-} ions then it is called caustic alkalinity.

1.10 DETERMINATION OF ALKALINITY

- Method
- Volumetric Method

Principle

The alkalinity of a given water sample is measured volumetrically by titrating with standard acids, e.g. HCl, H_2SO_4 and HNO_3. Using two different indicators titration carries out, one is phenolphthalein and the other one is Methyl orange. Phenolphthalein indicator is used to measure the hydroxide and carbonate alkalinity. Colour change from pink to colourless at neutralization end point. From this end point total carbonate ions get converted to bicarbonates.

This alkalinity is called phenolphthalein alkalinity. This end point is called phenolphthalein end point.

It is mentioned as "P"

Beyond this end point the same solution is used to titrate by using Methyl orange indicator.This end point explains the Bicarbonate alkalinity. Colour changes from pale yellow to pink colour. The total volume of acid consumed for this is called Methyl orange alkalinity. This end point is the complete neutralization of alkalinity causing anions.

It is mentioned as "M"

Phenolphthalein indicator changing the following conversion

$$OH^- + H^+ \longrightarrow H_2O$$
$$CO_3^{2-} + H^+ \longrightarrow HCO_3$$

Methyl orange indicator changing the following

$$HCO_3^{2-} + H^+ \longrightarrow H_2O + CO_2$$

Types (combination) of alkalinity anions in water

1. Only OH^- ion (hydroxide)
2. Only CO_3^{2-} ion (carbonate)
3. Only HCO_3^- ion (bicarbonate)
4. Combination of OH^- and CO_3^{2-} ions
5. Combination of OH^- and HCO_3^- ions

These are all possible types of alkalinity.

The other combination of hydroxide and bicarbonate ions are impossible to be in the same water.

Reason for the Above Statement

They are reacting together and forming water and carbonates.so carbonate alkalinity increases in water.

Reaction

$$OH^- + HCO_3^- \longrightarrow H_2O + CO_3^{2-}$$
$$NaOH + NaHCO_3 \longrightarrow H_2O + Na_2CO_3$$

From the above reaction we came to the conclusion, i.e. all the three alkalinity causing anions do not exist in the same water sample.

Procedure to Calculate Alkalinity

100 ml of the given water sample is pipetted out into a clean conical flask **Phenolphthalein** indicator added to the water sample (Pink colour formed). Then the water sample is titrated against sulphuric acid HCl. At end of the neutralization reaction pink colour gets converted to a colourless solution.Volume of acid is mentioned as 'P' ml.

Then to the same titrated water sample, 2 to 3 drops of **Methyl orange** indicator is added. Initial colour is pale yellow. The solution is titrated against the standard sulphuric acid. End point is that colour changes from yellow to pink. This is called 'M' ml.

"P" ml represents OH^- and CO_3^{2-} to HCO_3^- reaction

"M" ml represents completion of total alkalinity.

Calculation

Based on the volume of "P" and "M" values, types of alkalinity is justified.

$$OH^- + \tfrac{1}{2}CO_3^{2-} = P \qquad \ldots 1$$
$$OH^- + CO_3^{2-} + HCO_3- = M \qquad \ldots 2$$

Eq(2) – Eq(1)

$$\tfrac{1}{2}CO_3^{2-} + HCO_3^- = M - P \qquad \ldots 3$$

1. If P = O; OH-, CO_3^{2-} are Absent and HCO_3^- Present.
2. If P = ½ M; CO_3^{2-} Present and OH-, HCO_3^- are absent.
3. If P = M; CO_3^{2-}, HCO_3 are absent. OH^- ion present
4. If P > ½ M; CO_3^{2-}, OH^- are present and HCO_3^- absent

$$OH^- + 1CO_3^{2-} = M \quad \text{... 4}$$

$$\tfrac{1}{2}CO_3^{2-} = M - P \quad \text{... 5}$$

Alkalinity only because of CO_3^{2-} ion is

$$CO_3^{2-} = 2(M - P) \quad \text{... 6}$$

Calculation of OH^- alkalinity

$$OH^- + CO_3^{2-} = M \quad \text{... 7}$$

$$\text{Sub } CO_3^{2-} \text{ value} = 2(M - P) \text{ in equation} \quad \text{... 7}$$

$$OH^- + 2(M - P) = M \quad \text{... 8}$$

$$OH^- = M - 2(M - P)$$

$$= M - 2M + 2P$$

$$OH^- = 2P - M$$

$$OH^- + \tfrac{1}{2}CO_3^{2-} = P \quad \text{... 1}$$

$$OH^- + HCO_3^- + CO_3^{2-} = M \quad \text{... 2}$$

If hydroxide ion is absent

$$OH^- = 0$$

Then, Sub OH = 0 in (1) and (2)

$$\tfrac{1}{2}CO_3^{2-} = P \quad \text{... 3}$$

$$CO_3^{2-} + HCO_3^- = M \quad \text{... 4}$$

$$CO_3^{2-} = 2P \quad \text{... 5}$$

Sub (5) in (4)

$$HCO_3^- + 2P = M \quad \text{... 6}$$

$$HCO_3^- = M - 2P$$

Titration values are based on M and P. The alkalinity causing anions are summarized.

Titration Volume	*OH^-*	*CO_3^{2-}*	*HCO_3^-*
P = 0	0	0	M
P = M	✓	×	×
P = M / 2	×	✓	×
P > M / 2	✓	✓	×
P < M / 2	×	✓	✓

Individual alkalinity of water sample can be determined by the following equation:

For specified or particular alkalinity

$$= \frac{\text{volume of acid consumed} \times \text{normality or strength of acid} \times 50 \times 100}{\text{volume of water sample}}$$

= Strength or Normality of the water sample × 50 × 1000 ppm

Titration (1) Phenolphthalein volume

Sulphuric acid consumed	V_1	= ... ml
Normality or strength of Acid	N_1	= N/50 = 0.02 N
Volume of Water Sample	V_2	= 100 ml
Water Strength	N_2	= V_1N_1/V_2
	N_2	= $V_1/100 \times 0.02$

Alkalinity equivalent to $CaCO_3$ is

$= N_2 \times$ Equivalent weight of $CaCO_3$ $\times$ 1000 ppm

$= V_1/100 \times 0.02 \times 50 \times 1000$

P = 10 V1 ppm

Titration (2) *Methyl Orange Volume*

Volume of H_2SO_4 $V_1 = (V_1 + V_2)$

Strength $N_1 = 0.02$ N

Volume of H_2O $V_2 = V_1N_1/V_2$

$= (V_1 + V_2) \times 0.02/100$

Strength $N_2 = \frac{V_1N_1}{V_2} = \frac{(V_1 + V_2) \times 0.02}{100}$

$$= \frac{(V_1 + V_2) \times 0.02 \times 50 \times 1000 \, CaCO_3}{100 \text{ equivalent}}$$

$M = 10(V_1 + V_2)$ ppm

Alkalinity of water sample was calculated by the volumetric titration method.

1.11 HARDNESS

Water contains different salts. Because of these salts, It does not produce lather (bubbles/frothing) with soap solution. This is called hardness of water. Due to the hardness, soap precipitates in water.

soap + water ⟶ not giving lather bubbles

Example, ***Salts:*** Ca, Mg

1.12 SOFTNESS

Pure water readily gives lather with soap. if salts are removed from water, it will produce Lather (bubbles) with Soap Solution. This water is called soft water.

Soap + water ⟶ Giving lather (bubbles)

Reason: Absence of Ca and Mg salts

Soap

Soap is potassium (K) and Sodium (Na) salts of higher fatty acids.
Example, Stearic, Oleic and Palmitic acid

Hardness Explanation with Reaction

Soap is a sodium stearate or sodium salt of higher fatty acids. It is soluble in water.

Hard water contains Ca and Mg salts which are not soluble in water. Because of this insoluble precipitates appear thus, water does not produce lather with soap. This property is called hardness.

Reaction

$$\underset{\text{Soluble soap}}{C_{17}H_{35}COONa} + CaCl_2 \longrightarrow \underset{\text{Insoluble soap}}{(C_{17}H_{35}COO)_2Ca} + 2NaCl$$

Detection of Hardness

1. The water produces white precipitate with soap solution
2. Hard water gives wine red colour with Eriochrome Black - T at pH = 8 – 10

1.13 TYPES OF HARDNESS

Based upon the salts present in water and anion nature, hardness is classified into two types:

a. Temporary hardness (or) Carbonate hardness (or) Weak alkaline hardness
b. Permanent (or) Non-carbonate hardness

Temporary Hardness

This hardness is because of Calcium (Ca) and Magnesium (Mg) Bicarbonate salts present in water. Heavy metal salts also cause temporary hardness. This hardness can be removed easily.

The simple process is boiling of water

Salts: $Ca(HCO_3)_2$ and $Mg(HCO_3)_2$

Temporary hardness removing methods

Water is boiled up to its boiling point. While heating bicarbonates are converted to carbonates salts. Compare to solubility of Bicarbonate. Bicarbonate is readily soluble than carbonate. Carbonate salts are precipitates.

Reaction

$$Ca(HCO_3)_2 \xrightarrow{\Delta} CaCO_3\downarrow + H_2O + CO_2\uparrow$$

$$Mg(HCO_3)_2 \xrightarrow{\Delta} MgCO_3\downarrow + H_2O + CO_2\uparrow$$

By using simple methods, hardness producing salts are removed. Then it is called temporary hardness. This hardness because of bicarbonate salts, that's why it is called carbonate hardness.

Permanent Hardness or Non-carbonate Salts

Permanent hardness because of Calcium and Magnesium chloride and Sulphate salts.

It can not be removed by using a simple method. These salts are highly soluble in water.

Salts: $CaCl_2$, $CaSO_4$, $MgCl_2$ and $MgSO_4$

Temporary Hardness	Permanent Hardness
1. It is due to Bicarbonate Salts	It is due to chloride sulphate salts
2. It can removed easily by boiling the water	It can not be removed easily
3. While removing precipitates are formed	No reaction
4. Medium soluble salts	Highly soluble salts
5. Alkaline hardness	Non - Alkaline hardness $CaCl_2$ and $MgCl_2$

1.14 TEMPORARY HARDNESS REMOVAL

i. Heating the water to its boiling temperature

$$Mg(HCO_3)_2 \xrightarrow{\Delta} MgCO_3\downarrow + H_2O + CO_2\uparrow$$

ii. Water treated with lime $(Ca(OH)_2)$

$$\underset{}{Mg(HCO_3)_2} + \underset{\text{lime}}{2Ca(OH)_2} \xrightarrow{\Delta} Mg(OH)_2\downarrow + 2CaCO_3\downarrow + 2H_2O$$

1.15 PERMANENT HARDNESS REMOVAL

i. Water treated with Na_2CO_3 (Soda)

$$CaCl_2 + \underset{\text{Soda}}{Na_2CO_3} \longrightarrow CaCO_3^{-} + 2NaCl$$

ii. Water passed through (Na_2Ze) Sodium Zeolite bed

$$CaSO_4 + Na_2Ze \longrightarrow CaZe + Na_2SO_4$$

$$MgCl_2 + Na_2Ze \longrightarrow MgZe + 2NaCl$$

Total Hardness

The summation of Temporary Hardness and permanent hardness is called Total Hardness.

Total Hardness = Temporary Hardness + Permanent Hardness

Units of Hardness

i. Milligrams in one lit (Mg/lit)

ii. Parts per million (ppm)

$$1 \text{ Million} = 10^6 = 1000000$$

$$1 \text{ kg} = 1000 \text{ gm}$$

$$N \text{ gm} = 1000 \text{ mg}$$

iii. 1 kg in n litre water

$$1\ kg = 1000\ gm$$

$$\text{To convert milligram} = 1000 \times 1000\ mg = 10^6\ mgs = 10^6\ \text{million}$$

$$1\ mg/lit = 1\ ppm\ \text{in 1 litre}$$

$$1\ mg = 1\ ppm$$

Other units are Clarkes Degree (°C) and French Degree (°Fr).

Hardness-based on Calcium Carbonate Equivalents

1. Hardness producing salts do not have round figure Molecular weights For an example: 50,100,200... Like values
2. All other salts are freely soluble in water

To rectify the above problems $CaCO_3$ is used to express the hardness.

1. It is easily available salt
2. Molecular weight is = 100 (Round whole number)
3. Equivalent weight is = 50 (Round whole number)
4. It is not freely soluble in water

Amount equivalent to $CaCO_3$ (Based on Molecular weight)

$$= \frac{\text{weight of hardness producing salt} \times 100 \ \text{amount equivalent to } CaCO_3}{\text{molecular weight of hardness producing salt}}$$

Amount equivalent to $CaCO_3$ (Based on equivalent weight)

$$= \frac{\text{weight of hardness producing salt} \times 50 \ \text{amount equivalent to } CaCO_3}{\text{equivalent molecular weight of hardness producing salt}}$$

If the weight of $CaCl_2$ is 40 gm/lit

$$\text{Then hardness amount equivalent to } CaCO_3 = \frac{40 \times 100}{111}\ gm/lit$$

1.16 HARDNESS ESTIMATION BY EDTA METHOD

Method: Volumetric titration method

Chemicals: Eriochrome Black - T (EBT), Ethylene Diamine Tetra Acetic Acid (EDTA) and Standard $CaCO_3$ Solution

Objectives: In water, sample amount of hardness present is estimated by complex metric titration.

EBT — Complex Agent (1)
EDTA — Complex Agent (2)

EDTA Character

i. It is insoluble in water
ii. It is a complexing agent
i. For hardness titration disodium salt is used for solubility

Structure

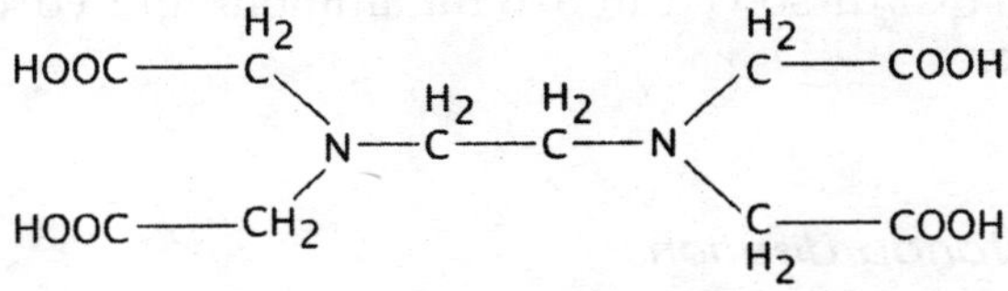

Principle

The total hardness producing salts present in water is estimated by titrating the hard water sample against EDTA and Eriochrome Black–T indicator at pH 8 – 10. To maintain pH 8 – 10. Buffer solution is used. Buffer containing $NH_4Cl + NH_4OH$ together.

Use of Buffer Solution

It maintains pH of the solution constant at all conditions.

From this titration, EBT forms a complex at this particular pH only.

Cations are quantitatively determined(Ca^{2+}, Mg^{2+}).

Reaction

Water containing different salts in it. In solution salts are in an ionic state.

So,

$$\text{Metal salt} \longrightarrow \text{Metal ion}$$

$$M \longrightarrow M^{2+} \quad \ldots 1$$

$$M^{2+} + EBT \xrightarrow{pH\,8-10} [M - EBT]\ \text{complex} \quad \ldots 2$$

Steel blue — Wine red (weak complex)

$$[M - EBT] + EBT \xrightarrow{pH\,8-10} [M - EDTA] + EBT \quad \ldots 3$$

Wine red — Steel blue (stable complex)

The first water sample is treated with Buffer and followed by 3 to 4 drops of EBT are added. Wine red colour forms. Then the solution is titrated against EDTA solution. When the wine red colour changes from steel blue that is the end point.

Different steps involved in this determination

1. Preparation of EDTA solution (Burette)
   a. 3.9 gm of solid Disodium salt of EDTA is dissolved in 1 litre of distilled water.
2. Preparation of EBT (Indicator)
   a. 0.5 gm of Eriochrome Blact – T solid is dissolved in 100 ml of Ethyl alcohol.
3. Preparation of standard Hard water (Pipette)
   a. 1 gm of pure and dry $CaCO_3$ is dissolved in minimum amount of dilute HCl and Made up to 1 litre solution.
   b. 1 gm = 1000 mg
   c. In 1000 ml of water, 1 gram of calcium carbonate is dissolved.
   In 1 ml of water, 1 mg of $CaCO_3$ is present

4. Preparation of the buffer solution
   a. 67.5 gm of NH_4Cl dissolved in 570 ml ammonium hydroxide and 430 ml of distilled water.

Procedure

Titration - I EDTA standardisation

Burette rinsed and filled by EDTA solution. Pipette out 50 ml of standard hard water ($CaCO_3$) solution into conical flask add 10 – 15 ml of buffer solution and followed by 3 to 4 drops of EBT indicator. Colour changes from wine red to steel blue. Titrate against EDTA. Volume consumed for colour change from wine red to steel blue is measured "V1" ml.

Titration - II Determination of Total Hardne

Pipette out 50 ml of the given Water sample into a clean conical flask Add 10 – 15 ml of buffer solution followed by addition of 3 to 5 drops of EBT indicator. Titrate against EDTA solution. End point is noted. Colour changes from wine red to steel blue and is:

Volume taken as "V2" ml.

Titration - III Determination of Permanent Hardness

A 250 ml water sample is taken in a beaker. Heated up to the boiling temperature of water for 15 minutes. Coolled and filtered through a funnel. While heating temporary hardness is removed. Then Pipette out 50 ml of the given water sample into a clean conical flask. Add 10 – 15 ml of buffer solution followed by addition 3 to 5 drops of EBT indicator. Titrate against EDTA solution. End point is noted. Colour changes from wine red to steel blue and is noted.

Then the same procedure is followed from the above

End point taken as "V3" ml.

Calculation

Titration - I Standardisation

1 ml of standard water containing 1 mg of $CaCO_3$

50 ml Standard water containing = 50 mg of $CaCO_3$

50 ml standard water calculated = V_1 ml of EDTA

V_1 ml of EDTA = 50 mg of $CaCO_3$

1 ml of EDTA = 50 /V_1 mg of $CaCO_3$

Titration - II Unknown (Total hardness)

50 ml of water sample = V2 ml of EDTA

Unknown EDTA = Standard EDTA

50 ml of Water = $V_2 \times 50/V_1$ mg of $CaCO_3$

1 ml of water = $V_2/50 \times 50/V_1$ ml of $CaCO_3$

= V_2/V_1 mg of $CaCO_3$ equivalent

Total hardness in

$$1 \text{ litre (or) } 1000 \text{ ml of } H_2O \text{ sample} = \frac{V_2 \times 100}{V_1} \text{ mg of } CaCO_3 \text{ eq}$$

$$\text{Total hardness} = Y \text{ g}$$

Titration - III Unknown Permanent Hardness

$$50 \text{ ml of boiled water} = V_3 \text{ ml of EDTA}$$

Sub EDTA from titration (1)

$$\text{In } 50 \text{ ml water} = \frac{V_3}{V_1} \times 50 \text{ mg of } CaCO_3 \text{ eq}$$

$$\text{In } 1 \text{ ml} = \frac{V_3 \times 50}{V_1 \times 50} \text{ mg of } CaCO_3 \text{ eq}$$

$$1 \text{ ml} = \frac{V_3}{V_1} \text{ mg of } CaCO_3 \text{ eq}$$

$$n\ 1000 \text{ ml (or) lit} = \frac{V_3}{V_1} \times 1000 \text{ mg of } CaCO_3 \text{ eq}$$

$$= Zg$$

$$\text{Total Hardness} = V_2/V_1 \times 1000 = Y \text{ gm}$$

$$\text{Permanent hardness} = V_3/V_1 \times 1000 = Z \text{ gm}$$

$$\text{Temporary hardness} = Y - Z \text{ gm} = X \text{ gm}$$

Boiler feed water and Troubles in Boiler Feed Water

1.17 BOILER FEED WATER

The water supplied or filled in boiler for producing steam is called boiler feed water.

Steam: water converted from liquid to vapour at a high pressure.

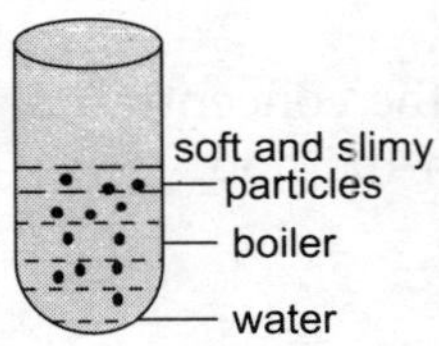

Boilers

i. In chemical industries
ii. Rice boilers in hotels, etc.
iii. Water heaters
iv. Modern rice mills
v. Steam boilers

Fig. 1.1

Characters of boiler feed water

i. Water should be free from turbidity

ii. Oil, Dissolved gases, Alkali producing salts, Hardness producing salts should be removed.

1.18 TROUBLES IN BOILER BECAUSE OF WATER

If the water is directly subjected to boiler without removal of impurities, it will be affecting boiler by following methods:

i. Scale and sludge formation

ii. Priming and Foaming (cary over)

iii. Caustic Embrittlement

iv. Boiler Corrosion

1.19 SLUDGE FORMATION IN BOILER

Example for Sludge

While Raining time the water looking brown in colour

Sludge Characters

The slimy soft and loose particles in water is called sludge. It is mingled in water.

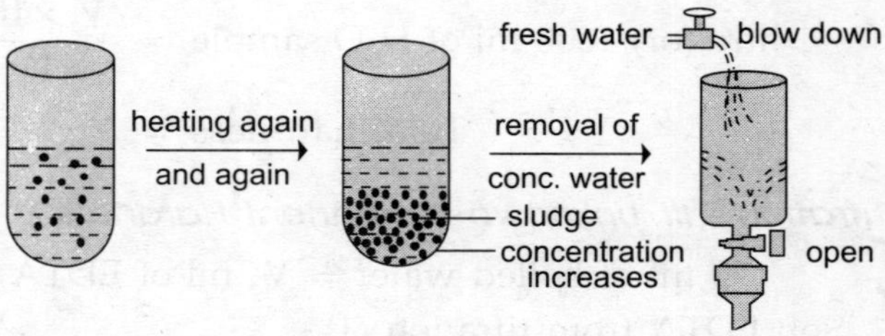

Fig. 1.2

Reason for sludge formation

The Boilers are heated again and again with water. Because of this continuous process, the salts concentration is increased in Boiler water. These salts are highly soluble in hot water and sparingly soluble in cold water.

Example, $MgCO_3$, $CaCl_2$, $MgCl_2$ and $MgSO_4$

Disadvantages

1. If Sludges are increased in water boilers. It will cause scale formation.
2. Because of the sludge particles heat energy wasted.
3. Boiler efficiency decreases
4. Steam temperature decreases and required energy increases.

Sludge Removal Methods

1. Sludge can be removed by using soft water
2. It is removed by resin purification or Membrane filtration of Boiler feed water
3. It can be removed by blow down method.

Blow Down Method

After steam preparation the concentrated water removed by fresh water every time. This is called blow down method.

Scale

Scale Examples

1. In water boilers or water heaters inside thick solid layer is formed.
2. Water heater surface is covered by thick solid (Scale).

Scale: The precipitates form a solid layer inside the boiler. It will become a hard layer while heating. This hard layer is called scale.

Example, ***Salts:*** $Ca(HCO_3)_2$, $CaSO_4$ and $Mg(OH)_2$

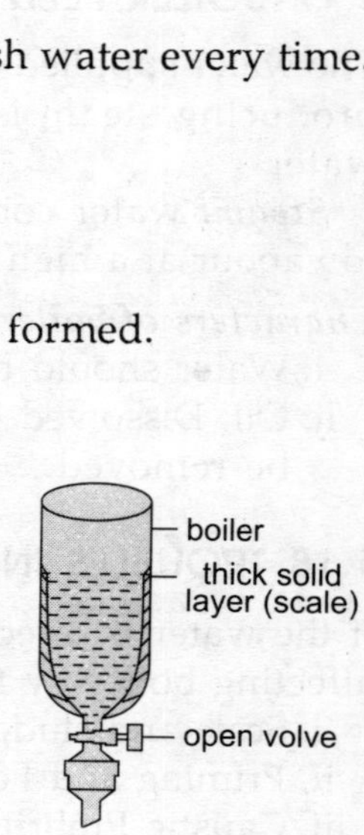

Fig. 1.3

Reasons for Scale

1. **Bicarbonate Decomposition**

$$\underset{\text{Soft particle}}{Mg(HCO_3)_2} \xrightarrow{\Delta} \underset{\text{Precipitate}}{MgCO_3\downarrow} + H_2O + CO_2\uparrow$$

$$MgCO_3 + H_2O \longrightarrow Mg(OH)_2 + CO_2\uparrow$$

Because of continuous heating, soft particles are converted to hard particles. They adhere on the inner boiler surface. Then it become scale inside the boiler. It forms only in low pressure boilers. But $CaCO_3$, $MgCO_3$ salts are soluble in high pressure boilers.

2. **Calcium Sulphate Deposition**
Calcium sulphate or Magnesium sulphate salt is soluble in cold water but insoluble in very hot water. The rate of deposition increases with increase in temperature. Temperature increases, solubility of $CaSO_4$ decreases.

Weight of $CaSO_4$	***Temperature***
3200 ppm	15°C
55 ppm	230°C
27 ppm	320°C

3. **Hydrolysis of Magnesium Salts**
Magnesium salts react with water. Salts are hydrolyzed and form soft Scales inside the boiler.

$$MgCl_2 + H_2O \longrightarrow \underset{\text{Precipitate}}{Mg(OH)_2\downarrow} + \underset{\text{Gas}}{2HCl\uparrow}$$

4. **Silica or Alumina Particles**
Silica forms magnesium or calcium silicates with salts. It is hard in nature so the silicate scale is formed inside the boiler. It is highly difficult to remove it from the boiler.

Example, $MgSiO_3$

Disadvantages of scale formation

1. Efficiency of boiler decreases
2. Scales develop cracks inside the boiler
3. Cracks cause accidents in plants.
4. Because of scales, heat energy is wasted
5. Steam supply is also affected.

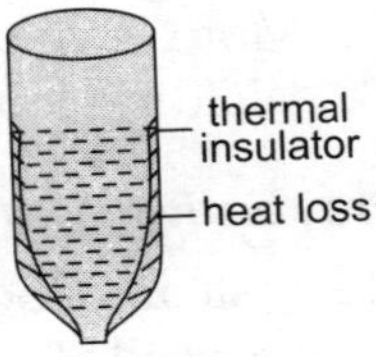

Fig. 1.4

Scale Removal Methods

1. Soft Scales are removed using wire brush by Scraping.
2. Boiler heated well and suddenly washed by cold water. Because of this process Scale gets broken. This process is called Thermal Shock.
3. Using some chemicals (Acids or Complexing agents)
4. **Example,** $CaCO_3$ removed by 1·0% HCl solution
   i. $CaSO_4$ removed by EDTA

Differences between Sludge and Scale

	Sludge	*Scale*
1.	Soft and floating precipitates	Hard and thick layer inside the boiler
2.	It does not cause any accident in boiler	It is causing accidents in boiler
3.	Cracking does not occur	Cracking occurs
4.	It is the easily removable one	It is difficult to remove
5.	Example salts: $MgCl_2$ and $CaCl_2$	$CaSO_4$ and $Mg(OH)_2$

1.20 CAUSTIC EMBRITTLEMENT

1. It is formed due to high alkalinity present in boiler feed water.
2. Caustic means alkalinity.
3. Due to alkalinity, corrosion occurs in the boiler.
4. Caustic-Embrittlement is inter crystalline breaking.

Reaction and Reason

1. While the softening process, lime soda is added.
2. Due to the above process, Na_2CO_3 is increased in water.

$$Na_2CO_3 + H_2O \longrightarrow 2NaOH + CO_2$$

This causes electrochemical corrosion in (Fe) iron boilers.

$$Fe + 2NaOH \longrightarrow Na_2FeO_2 + H_2\uparrow$$

[Na_2FeO_2—Sodium Ferroate]

Sodium Ferroate is dissolved in water. Because of this product, cracks are formed in the boiler.

Reasons

1. Wrong designing of boilers
2. More Rivets nail used in boilers
3. Welding joints and bends

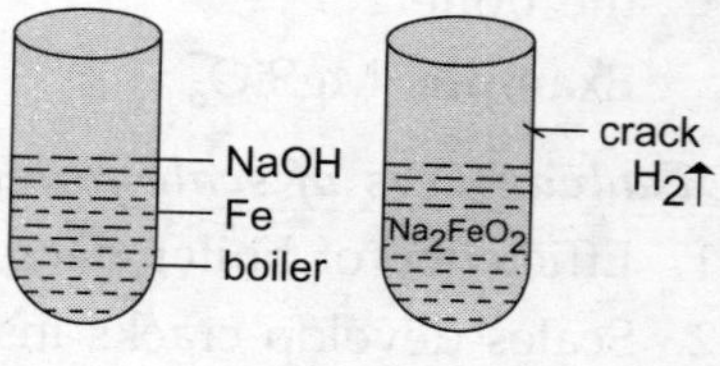

Fig. 1.5

Preventive Methods

1. Instead of Sodium carbonate for the softening process, Sodium phosphate should be used.
2. To the boiler we have to use lignin. Pectin with $MgSO_4$.
3. These chemicals block the cracks inside the boiler.

1.21 PRIMING AND FOAMING

Steam: It is the vapour form of water with higher pressure and temperature.

Wet steam: During the preparation of steam, the boiler is heated rapidly. At this time (steam), vapours carry water droplets with themselves.

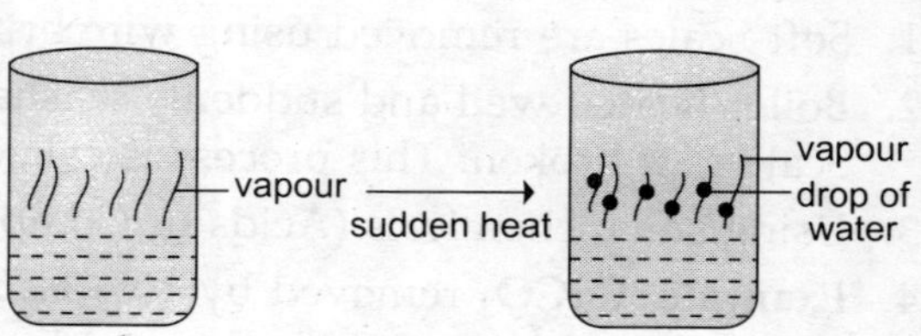

Fig. 1.6

Carry Over

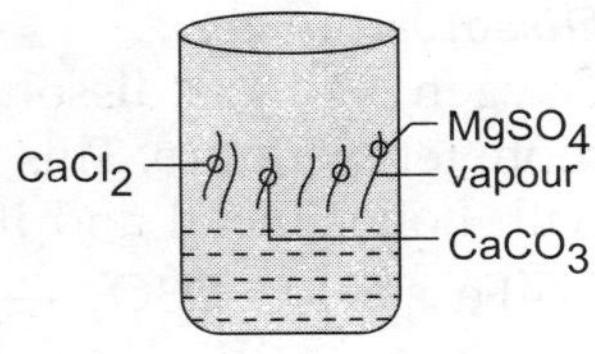

Fig. 1.7

While formation of wet steam water droplets taking suspended particles or dissolved salts. This is called carry over.

This carry over process occurs due to Priming. Priming is the process of wet steam preparation.

Reason for Priming

Because of hardness producing salts and dissolved solids, sudden heating of boiler and higher steam velocities cause priming in boiles. While filling a boiler, the water level should be in limit otherwise it causes Priming.

Remedies

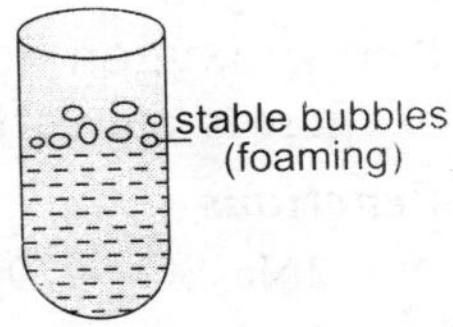

Fig. 1.8

1. Avoid sudden boiling
2. Proper boiler design
3. Always use pure water/treated water
4. The water level must be maintained inside the boiler.

Foaming

Formation of stable bubbles inside the boiler over the surface of water is called Foaming.

If bubbles are foaming more and more it causes priming inside the boiler.

Reason: Because of floating particles inside the water

Example, oil, grease and Micro particles

Remedies

i. Adding coagulants with water

Example, Sodium aluminate and Aluminum hydroxide

Action of Coagulants

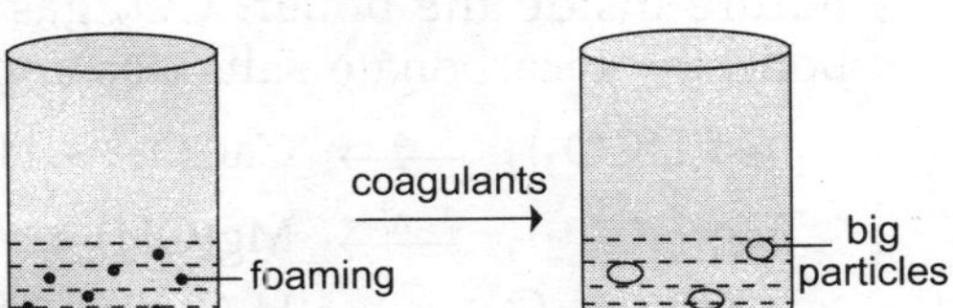

Fig. 1.9

Small particles are converted to big particles and precipitated. Anti foaming agents are added to the water

Example, Castor oil or Synthetic polyamides.

All these convert and precipitate floating particles.

1.22 BOILER CORROSION

Boiler corrosion is the gradual degradation of boiler materials by dissolved chemicals of water.

Example for dissolved chemicals–O_2, CO_2 gas and salts.

Dissolved oxygen

Oxygen (O_2) gas dissolved in water. Maximum limit of oxygen is 8 mg per liter at room temperature. This O_2 attacks boiler inside at higher temperature. Oxygen reacts with boiler metal and it forms iron hydroxide.

$$4Fe + 6H_2O + 3O_2 \longrightarrow 4Fe(OH)_3$$

Removal of Dissolved Oxygen

Based on the removal methods it is classified into two types:

I. Chemical Method
II. Physical Method

Chemical Method

Using oxygen reacting chemicals, dissolved oxygen is converted as oxygenated products.

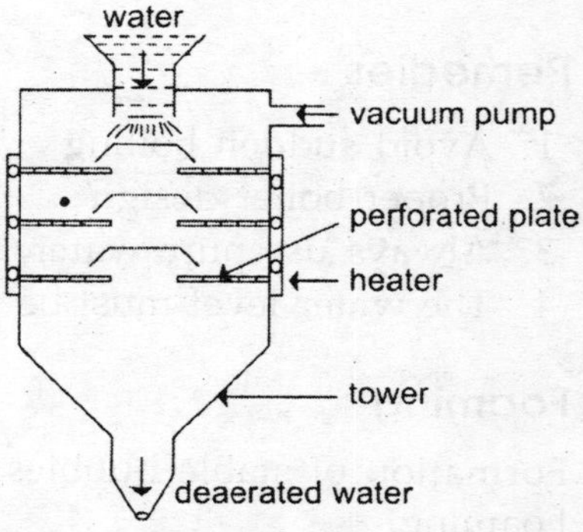

Fig. 1.10

Reactions

$$2Na_2SO_3 + O_2 \longrightarrow 2Na_2SO_4$$

Sodium sulphite → Sodium sulphate

Good Chemical

$$N_2H_4 + O_2 \longrightarrow N_2 + 2H_2O$$

Hydrazene

$$Na_2S + 2O_2 \longrightarrow Na_2SO_4$$

Physical Method

Removal of air (O_2) from water is called deaeration. Mechanical deaeration is removal of air by mechanical energy.

1. **Instrumentation:** A machine is connected with vacuum pump and it is used to suck oxygen. Steam jacket heating the water, then O_2 gas evolves. Temperature increases and pressure decreases inside the tower.
2. **Dissolved carbon dioxide:** Dissolved carbon dioxide in water produces acidic nature inside the boiler. CO_2 gas forms carbonic acid with water inside the boiler by bicarbonate salts decomposition.

$$Ca(HCO_3)_2 \xrightarrow{\Delta} CaCO_3\downarrow + H_2O + CO_2\uparrow$$

$$Mg(HCO_3)_2 \xrightarrow{\Delta} Mg(OH)_2\downarrow + CO_2\uparrow$$

$$CO_2 + H_2O \longrightarrow H_2CO_3 \text{ (Carbonic acid)}$$

Remedies or Prevention Method

a. CO_2 removed by adding quantified ammonia into water

$$NH_3 + H_2O \longrightarrow NH_4OH$$

$$2NH_4OH + H_2CO_3 \longrightarrow (NH_4)_2CO_3 + 2H_2O$$

b. Mechanical – aeration process
c. Continuous floating of water

3. **Dissolved salts:** Dissolved permanent hardness producing salts cause acidity inside the boiler. The acidity causes boiler corrosion.

   Example, $MgCl_2$ and $CaCl_2$

   Hydrolysis at higher temperature

   $$CaCl_2 + 2H_2O \longrightarrow Ca(OH)_2\downarrow + 2HCl$$

   Acid reacts with Boiler Metals

   $$Fe + 2HCl \longrightarrow FeCl_2 + H_2\uparrow$$

   $$FeCl_2 + 2H_2O \longrightarrow Fe(OH)_2 + 2HCl$$

   From the above reactions, HCl is formed again and again.

Prevention Method

Acidity created by the salts is neutralized by sodium hydroxide. By the titration method (volumetric) NaOH quantity is calculated

$$HCl + NaOH \longrightarrow NaCl + H_2O$$

1.23 SOFTENING METHODS

Water is utilized by many industries for different purposes. So, it should be free from suspended particles, hardness producing salts, gases and impurities. Removal of hardness producing salts from raw water is called the softening process.

Based on the removal methods it is classified into two types

1. Internal Treatment
2. External Treatment

1.24 INTERNAL TREATMENTS

Softening process is carried or done after the water is filled inside the boiler

Boiler Compounds

Chemicals used for internal treatment process are called Boiler compounds. Boiler compounds convert insoluble carbonates, chlorides and sulphates of Ca and Mg salts into soluble or floating loose particles. Then, loose particles are removed by blow down operation.

Based on the chemicals used for this process, these are classified into six types:

1. Calgon conditioning
2. Phosphate conditioning
3. Carbonate conditioning
4. Colloidal conditioning
5. Sodium aluminate treatment
6. Electrical conditioning

Calgon Conditioning

Calgon is sodium hexa Meta phosphate $Na_2\,[Na_4\,(PO_3)_6]$. It reacts with hardness producing salts and forms a highly water soluble complex. It does not affect the boiler action.

$$Na_2\,[Na_4\,(PO_3)_6] + 2CaSO_4 \longrightarrow Na_2[Ca_2\,(PO_3)_6] + 2Na_2SO_4$$

Phosphate Conditioning

It is applicable in high pressure boilers. Hardness producing salts react with phosphate salts. These solid particles are converted to loose and soft particles. These particles are removed by blow down operation.

Example, Na_3PO_4 Salt

Different phosphate salts

Na_3PO_4 – Tri sodium phosphate (Highly Alkaline)

Na_2HPO_4 – Disodium hydrogen phosphate (Weak alkaline)

NaH_2PO_4 – Sodium dihydrogen phosphate (Acidic)

$$3CaSO_4 + 2Na_3PO_4 \longrightarrow Ca_3(PO_4)_2 + 3Na_2SO_4$$

Carbonate Conditioning

It is applicable for low-pressure boilers. Sodium carbonate is used for the removal of hardness producing salts. By this process, insoluble precipitates are formed. It is removed by the blow down method.

Example, Na_2CO_3

Reaction:

$$Na_2CO_3 + CaSO_4 \longrightarrow Na_2SO_4 + CaCO_3$$

Colloidal Conditioning

Colloidal agents are organic compounds.These chemicals or compounds cover the scale surface and get converted to loose solids.It is removed by blow - down operation.It is applicable for low pressure boilers

Example, kerosene, agar - agar gel, gelatin

Sodium Aluminate Treatment (Na_2AlO_2)

In this process, Sodium aluminate is directly added to the boiler water. By this addition, Sodium hydroxide (NaOH) is formed inside the boiler and [Al $(OH)_3$] aluminium hydroxide slimy gelatinous precipitate forms. Double action occurs in this process.

$$NaAlO_2 + 2H_2O \longrightarrow Al(OH)_3 + NaOH$$

$Al(OH)_3$ –Covering scale forming particles

NaOH reacts with scale forming particles

$$NaOH + MgCl_2 \longrightarrow Mg(OH)_2 + 2NaCl$$

These particles are removed by blow down operation.

Electrical Conditioning

Scaled glass bulbs containing mercury are connected to a battery and the bulbs are placed in the boiler water and kept rotating. Mercury bulbs produce electrical discharges. It prevents the scale forming solids. It effectively controls the formation of hard scales

1.25 EXTERNAL (CONDITIONING) TREATMENT

This process purifies water from hardness producing salts, before feeding into the boiler.

Based on chemicals and principles it is classified into three types

1. Lime soda process
2. Zeolite process
3. Ion exchange process

Lime Soda Process

Lime - $Ca(OH)_2$
Soda - Na_2CO_3

Principle

In this method, all the soluble hardness causing constituents get converted to insoluble precipitates Then it is removed.

Procedure

A suspension of calcium hydroxide (lime) and sodium carbonate (soda) is added to water in calculated quantities, based on the concentrations of impurities they are expected to remove and their nature and stoichiometry of chemical reactions.

Reactions of Lime

$$Ca(OH)_2 + CO_2 \longrightarrow CaCO3? + H_2O$$
$$Ca(OH)_2 + H_2S \longrightarrow CaS + 2H_2O$$
$$Ca(OH)_2 + 2HCl \longrightarrow CaCl_2 + 2H_2O$$
$$Ca(OH)_2 + Ca(HCO_3)_2 \longrightarrow 2CaCO_3\downarrow + 2H_2O$$
$$Ca(OH)_2 + Mg(HCO_3)_2 \longrightarrow Mg(OH)_2\downarrow + 2CaCO_3\downarrow + 2H_2O$$
$$Ca(OH)_2 + MgCl_2 \longrightarrow Mg(OH)_2\downarrow + CaCl_2$$
$$Ca(OH)_2 + FeSO_4 \longrightarrow Fe(OH)_2\downarrow + CaSO_4$$

Reaction of Soda

$$Na_2CO_3 + CaCl_2 \longrightarrow 2NaCl + CaCO_3\downarrow$$
$$Na_2CO_3 + MgSO_4 \longrightarrow MgCO_3\downarrow + Na_2SO_4$$

Water softened by this method contains considerable weight of soluble salts like NaCl, Na_2SO_4 present. So it can not be used in high pressure boilers.

Zeolite or Permutit Process

Zeolite — Greek meaning — boiling stone.
It is also known as Permutit.
Molecular formula - $Na_2O.Al_2O_3.XSiO_2.YH_2O$
Where $X = 2 - 10$; $Y = 2 - 6$

Principle

Zeolite holds sodium ions loosely and can be simply represented as Na_2Z and Z is the insoluble radical framework. By this process, only cations are removed during hard water passing through a bed.

Reactions

$$CaCl_2 + Na_2Z \longrightarrow CaZ + 2NaCl$$
$$MgCl_2 + Na_2Z \longrightarrow MgZ + 2NaCl$$
$$CaSO_4 + Na_2Z \longrightarrow CaZ + Na_2SO_4$$
$$Ca(HCO_3)_2 + Na_2Z \longrightarrow CaZ + 2NaHCO_3$$

- Eventually, the bed loses its sodium exchange capacity
- Then zeolite is said to be "exhausted"zeolite

Regeneration of zeolite

Zeolite is regenerated by washing the bed with a concentrated solution of Sodium chloride (Brine Solution)

$$CaZ + 2NaCl \longrightarrow CaCl_2 + Na_2Z$$
$$MgZ + 2NaCl \longrightarrow MgCl_2 + Na_2Z$$

Exhausted Zeolite Regenerated Zeolite

Residual hardness of water from this process is 0 – 15 ppm.

Ion Exchange Process

Phenomenon: Adsorption phenomenon

Ion exchange resin: It is an organic, insoluble polymer. Cross link bonds are present in it. Functional groups are responsible for the ion exchanging process.

Principle

Cation exchangers

It is a type of neutralization. Acid functional groups are capable of exchanging their H^+ ions with other cations which come into contact with them. They are termed as "cation exchangers"

Functional Groups: –COOH and $–SO_3H$

Cation exchange resins are

1. Styrene divinyl benzene co-polymers
2. On sulphonation or carboxylation, they acquire a capability to exchange their hydrogen ions.

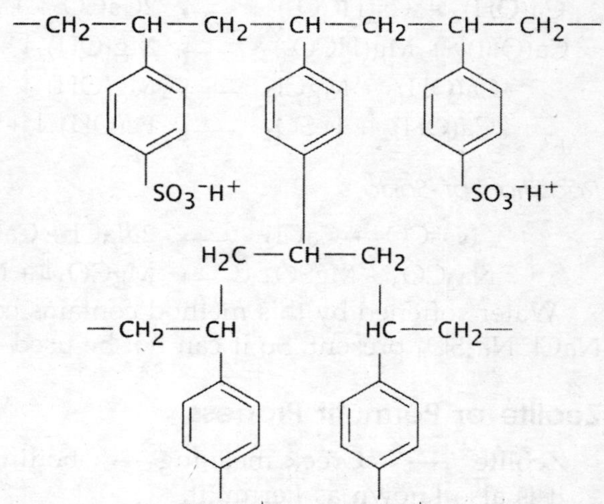

Cation exchange resin

Anion exchangers

Resin containing basic functional groups.Functional group such as Amine

Example, NH_2 and R_2NH

These functional groups are capable of exchanging their anions with other anions in water and therefore, these are known as "anion exchangers"

Anion exchange resins are Styrene divinyl benzene or Amine formaldehyde copolymers.

Molecular structure

This is simply mentioned as $\mathbf{R^-H^+}$

Reactions

$$2R^- H^+ + Ca^{2+} \longrightarrow R_2Ca + 2H^+$$

$$2R^- H^+ + Mg^{2+} \longrightarrow R_2Mg + 2H^+$$

Commercially available Resins

Amberlite IR – 120

Dowex – 50

Anion Exchange Resin

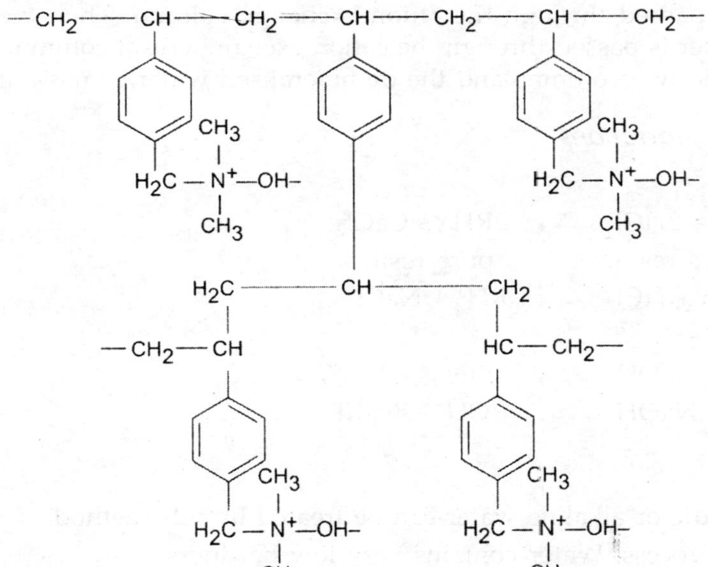

Simple form is R^+OH^-

Reactions

$$R^+ OH^- + Cl^- \longrightarrow RCl + OH^-$$

$$2R^+ OH^- + SO_4^{2-} \longrightarrow R_2SO_4 + 2OH^-$$

$$2R^+ OH^- + CO_3^{2-} \longrightarrow R_2CO_3 + 2OH^-$$

$$R^+ OH^- + HCO_3^- \longrightarrow RHCO_3 + OH^-$$

Commercially available anion resin

Amberlite – 400

Dowex – 3

This material (resin) is called Deioniser or Demineraliser. This process is called

Deionization or Demineralization

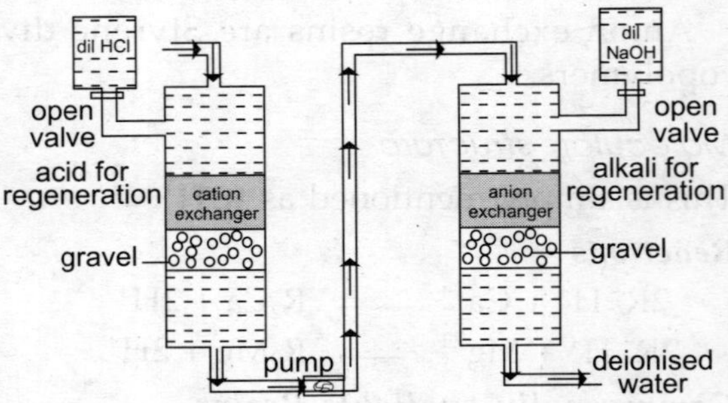

Fig. 1.11

Deionisation

By this process Anions (Cl^-, SO_4^{2-} and etc.) and Cations (Ca^{2+}, Mg^{2+} and etc.) are removed by resins cations, anions both are removed by this process. That's why it is called deionisation process.

Demineralisation

By this process, all minerals are removed ($CaCl_2$ and $MgSO_4$). Hence it is called demineralisation.

Process

Raw water is passed through the anion exchange column. Then the same cation containing water is passed through the cation exchange resin column. Both ions are removed by the two columns and the demineralised water comes out.

Regeneration Reactions

Cation resin

$$R_2Ca + 2HCl \longrightarrow 2RH^+ + CaCl_2$$

Impure resin — pure resin

$$RNa + HCl \longrightarrow RH^+ + NaCl$$

Anion resin

$$R_2SO_4 + 2NaOH \longrightarrow 2ROH^- + Na_2SO_4$$

$$RCl + NaOH \longrightarrow ROH^- + NaCl$$

Advantages

1. Highly acidic or alkaline water can be treated by this method
2. After the process, Water contains very low hardness
3. Dissolved solids are totally absent from this process.

Disadvantages

1. More than 10 ppm turbidity can not be used for the softening process
2. Equipment cost is very high
3. Fe, Mn containing water cannot be treated. These form stable complexes with resins.

From anion exchange column, the liberated H^+ ion and cation exchange column liberated OH^- ion. These ions react together and form water.

$$H^+ + OH^- \longrightarrow H_2O$$

The residual hardness from this method is 0 – 2 ppm

These Cation exchange resins are exhausted after the process. Then it is regenerated by using acid (HCl, H_2SO_4 diluted). Anion exchange resins are regenerated by Alkali (NaOH, KOH) wash.

Comparisons of Lime soda, Zeolite and ion exchange resin Methods:

No.	*Lime Soda*	*Zeolite*	*Ion Exchange*
1.	Residual hardness 15 – 50 ppm	Residual hardness 0 – 15 ppm	Residual hardness 0 – 2 ppm
2.	Acidic water can also be treated	Acidic water cannot be treated	Highly acidic and alkaline water can be treated
3.	Lesser amount of dissolved solids	More dissolved solids	No dissolved solids
4.	Operating expenses are high	Operating expenses are low	Operating expenses high

No.	*Lime Soda*	*Demineralisation*
1.	Lime $[Ca(OH)_2]$, Soda $[Na_2CO_3]$ are required	Cation and anion exchange resins are required
2.	Fe and Mn containing water can also be treated	It cannot be treated
3.	Turbid water can be treated	Turbid water cannot be treated
4.	Hardness producing salts are removed	All minerals are removed
5.	Residual hardness is high (15 – 50 ppm) zero is not possible	Zero is possible

No.	*Zeolite*	*Ion Exchange*
1.	Cations are only exchanged by this process	Both ions are exchanged
2.	Acidic water cannot be treated	Acidic and alkaline water are also treated
3.	Regeneration done by sodium chloride solution	Regeneration is done by HCl for Anion resin and by NaOH for cation resin
4.	Dissolved salts are more	No dissolved salts
5.	This is not used as boiler feed water	This can be used as boiler feed water
6.	Residual hardness 0 – 15 ppm	Residual hardness 0 – 2 ppm

Soft Water

1. It does not contain hardness producing Ca^{2+} and Mg^{2+} ions
2. It may contain Na^+, microorganisms.

Deionised Water

1. It does not contain any ions
2. It may contain some amount of organisms.

Distilled Water

1. It does not contain any ions and microorganisms
2. It is prepared by condensing the water vapour.

No.	External Treatment	Internal Treatment
1.	It is carried out before being filled into the boiler	After being filled into the boiler
2.	Preventive method	Corrective method
3.	It occurs in high pressure boiler	Low pressure boiler
4.	Zeolite process, ion exchange process	Carbonate, Phosphate conditioning

1.26 DESALINATION OF BRACKISH WATER

Brackish Water

Water containing high concentrations of dissolved solids is known as brackish water.

Example, Water present near the sea. It contains 3.5% of salts.

Desalination

Removal of common salts from water is called "Desalination"

Different Techniques

i. Distillation
ii. Electrodialysis
iii. Reverse osmosis

Based on dissolved salts, water types are as follows:

Fresh water

It has less than 1000 ppm of dissolved solids

Brackish Water

It has greater than 1000 ppm and less than 35000 ppm of dissolved solids

Sea Water

It has greater than 35000 ppm of dissolved solids.

Electrodialysis

Principle

By using current, negative (-ve) and positive (+ve) ions are pulled.

Process: Desalination

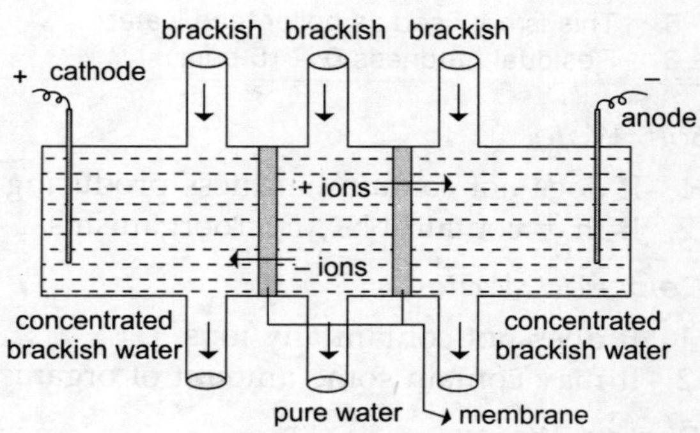

Fig. 1.12

Operation

Water is passed through the column. It contains electrodes and a set of ion selective membranes. While current is passed through the column, ions are pulled by electrodes. Pure water come at the centre. Brackish and concentrated impure water come out at sides.

Reverse Osmosis

Name of the Process: Desalination

Principle

When two solutions of different concentrations, are separated by a semi permeable membrane.

If pressure is applied to the solution, solvent flow from lower concentrations to higher concentrations is called osmosis. The pressure is called Osmotic pressure. If the solvent moves from higher concentration to lower that is called reverse osmosis.

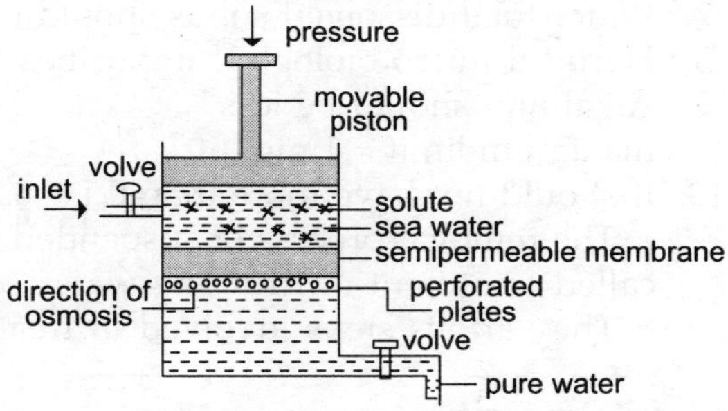

Fig. 1.13

Solution = Solvent + Solute

Brackish H_2O = Dissolved salts + Water

Semi permeable membrane allows only the solvent.

Procedure

Sea water is passed through the reverse osmosis column. The piston is moved forward. Brackish water gets pressure inside. Dissolved salts stay on semi permeable membrane. Then, pure water come out of the cylinder. Applied Pressure ranges from 15 to 40 kg/cm^2. This process is called Super filtration or hyper filtration.

Advantages

1. This process is cost effective.
2. It provides boiler feed water in high pressure boilers.
3. It removes ionic, non-ionic, colloidal and high molecular weight organic matters.

1.27 TREATMENT OF DOMESTIC OR POTABLE WATER

Drinking water or potable water should satisfy the following requirements before they are supplied (Potable Water - Drinking Water).

Important requirements of drinking water

Based on Salts and dissolved gases, the requirements made by BIS – 10500:1991 (International Standard Symbol)

1. It should be colourless and odourless (Hazam unit - max 5)
2. It should be of pleasant taste and odourless
3. It should be clear in nature
4. pH value should be 6.5 – 8
5. It should be free from dissolved gases like H_2S, CO_2 and NH_3
6. It should be free from toxic chemicals like

   Lead (Pb) – 0.05 mg/lit (max. limit)
   Arsenic (As) – 0.05 mg/lit (max.limit)
   Chromium (Cr) – 0.05 mg/lit (max.limit)
   Manganese (Mn) – 0.05 mg/lit (max.limit)

7. Water total dissolved solids should be less than 500 ppm
8. Harmful micro biological impurities like bacteria should be removed
9. Alkalinity should be less
   maximum limit = 1 mg/litre
10. It should not have more than 0.1 – 0.2 ppm of free chlorine

The removal process of suspended and colloidal impurities from raw water is called treatment of drinking water.

The various steps involved in treatment of water

Impurity	*Process used for removal*
1. Floating materials such as leaves and twigs	Screening
2. Suspended impurity such as sand, clay etc	Sedimentation
3. Fine suspended inorganic matter	Sedimentation with coagulation
4. Colloidal impurities and larger microorganisms	Filtration
5. Pathogenic bacteria	Disinfection or Sterilization
6. Maintain O_2 and removal of other gases	Aeration

Block Diagram

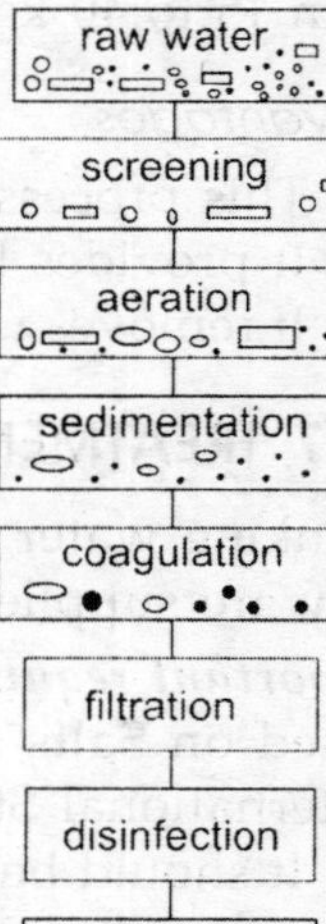

Fig. 1.14

Screening

The removal process of floating materials from the raw water is called Screening.

1. A screen contains a lot of holes in it
2. Water is passed through this
3. Leaves, wood some other floating particles are removed

Aeration

From the raw water, unwanted dissolved gases are removed by passing air through water which is called aeration.

Sedimentation

This process based on gravitational force. Up to 75% of suspended impurities are removed by this process. Water is allowed to stand for 4 – 6 hours in very big tanks without disturbance. Impurities settle down. The rate of sedimentation, is based on the Size, Specific gravity, Shape of the particle and temperature of water.

Coagulation

Finely divided impurities like silica, clay and organic matter are removed by this process.

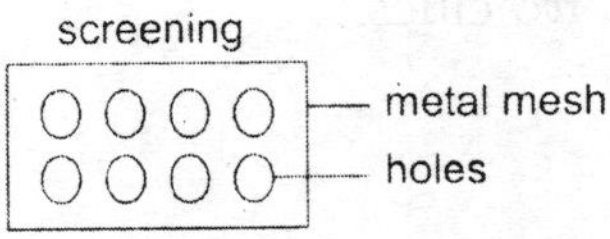

Fig. 1.15

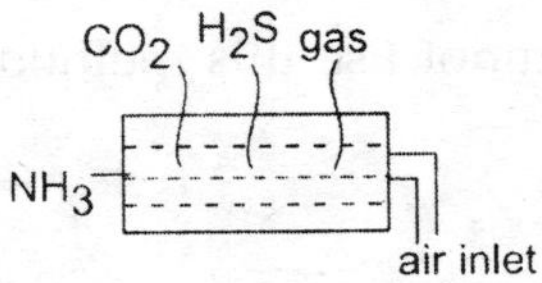

Fig. 1.16

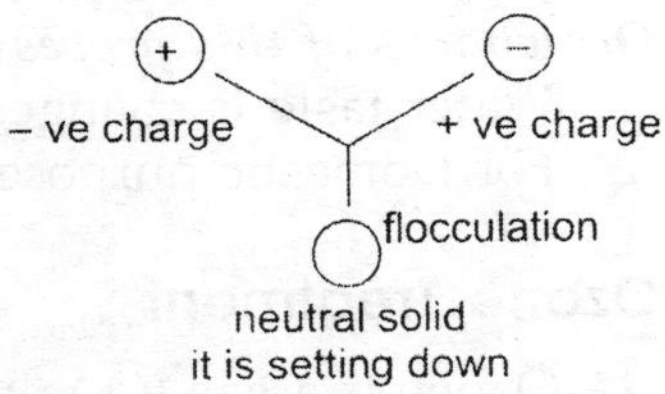

Fig. 1.17

These fine particles contain negatively charged species. Example for coagulating agents are Ferric sulphate, Ferrous sulphate, Cuprous chloride, Alum, Ammonium alum and Sodium aluminate.These are all called Coagulants.

Fine particle impurities have a negative charge and chemicals added to this have a positive charge. Neutralization takes place. Small particles are converted to big particles. These big particles settle down. Settling by chemicals is called flocculation. This total process is called coagulation. Chemicals used for this process is called Coagulants. Settled particles are removed easily by filtration.

Filtration

Water passing through filter beds. Filter bed contains fine, coarse, sand and fine coarse gravel.This process removes bacteria, colour, taste, odour and fine suspended particles.

Block Diagram of Filtration Process

Based on gravitational force, water is purified. Impurities fix on the surface

Surface is purified by scratching the top surface.

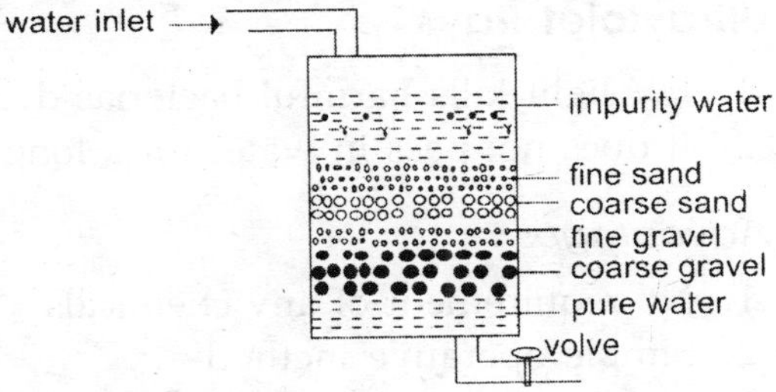

Fig. 1.18: Filtration

Sterilisation or Disinfection

The technology or process used to remove the biological impurities from water is called sterilisation. The chemicals used for the sterilization process are called Sterilisers or disinfectants.

Examples for Sterilisers: Chlorine gas, Bleaching power, Ultraviolet rays and Ozone gas.

Different Sterilisation Methods

1. Boiling Process
2. Chlorination
3. Ozone Treatment
4. Ultraviolet Process

Boiling Process

1. Water is heated up to its boiling point for 10 – 15 minutes
2. All biological impurities are totally destroyed.

Drawbacks of this process

1. Water taste is changed
2. For Domestic purpose we cannot use this method in Metro cities

Ozone Treatment

1. Ozone is a gas (O_3)
2. It is a powerful disinfectant
3. It readily reacts with water and forms nascent or atomic oxygen
4. Nascent oxygen destroys biological impurities like bacteria

$$O_3 \longrightarrow O_2 + [O]$$

$$[O] + \text{bacteria} \longrightarrow \text{bacterias killed}$$

Advantages

1. O_3 is a multi purpose agent. It is a sterilizing, bleaching, decolourising and deodourising agent.
2. It is not harmful in water.

Disadvantages

1. Its high cost
2. Storage of O_3 is a problem. Because it is unstable.

Ultraviolet Rays

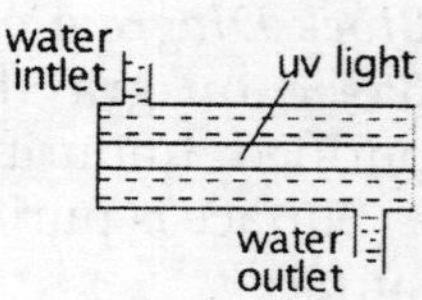

Fig. 1.19

1. UV light kills harmful bacterias during sterilization.
2. It does not pass in water for a long depth.

Advantages

1. No requirement of any chemicals
2. Simple operative method
3. No changes in colour or odour
4. It is used to clean water.

Disadvantages

1. Cost is very high
2. It should be handled carefully.

Chlorination

Addition of Chlorine gas to water is called Chlorination.

Chlorination by different methods

1. Direct Chlorine Gas (Cl_2)
2. Chloramines Addition ($ClNH_2$)
3. Bleaching Power ($CaOCl_2$).

Chlorine Gas Addition

1. In this process, Cl_2 gas is added directly to water. Harmful bacteria are destroyed.

Chloramine Addition

1. Chlorine and ammonia are added together in the ratio of 2 : 1
2. Chloramine are formed and added with water. It decomposes and the Cl_2 gas is slowly evolved and mixed with water.

$$\underset{2\,:\,1}{Cl_2 + NH_3} \longrightarrow \underset{\text{Chloramine}}{ClNH_2 + HCl}$$

Bleaching Powder ($CaOCl_2$)

1. Bleaching powder is calcium hypochloride
2. By producing nascent oxygens, hypochlorous acid, it acts as a sterilizing agent.
3. Hypochlorous acid corrodes the micro bacteria cell membrane

$$CaOCl_2 + H_2O \longrightarrow Ca(OH)_2 + Cl_2$$
$$Cl_2 + H_2O \longrightarrow HOCl + HCl$$
$$HOCl \longrightarrow HCl + [O]$$

Drawbacks

1. Stability of bleaching power is less
2. It increases hardness (Ca^{2+})
3. It produces smell and taste in water

Smell and taste are removed by adding ammonia.

Break point of chlorination

1. The action of chlorine gas in water is explained by the graph
2. Water has bacteria, organic impurities and reducing compounds (Fe^{2+})
3. At which particular concentration Cl_2 gas available for sterilisation is called break point of chlorination.

In graph,

0A : Initially applied chlorine gas utilized by reducing components of water. So initially Cl_2 gas is not available.

AB : Applied chlorine reacts with organic compounds here formation of chloro compound occurs.

BC : Decomposition of Chloro compounds up to the "C" point.

CD : Decomposed Cl_2 gas is involved in sterilisation

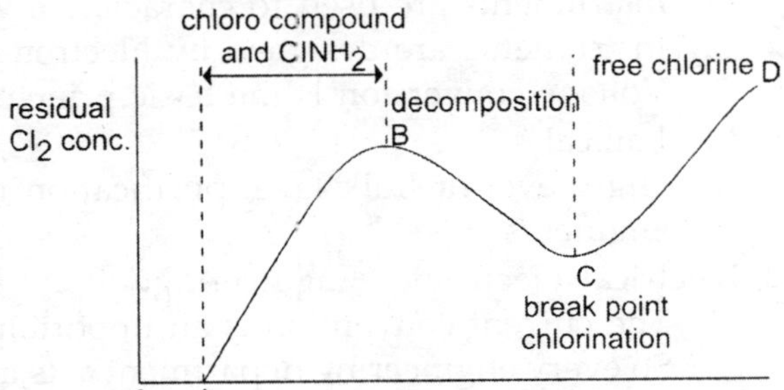

Fig. 1.20: Breakpoint of chlorination graphical representation

C : At this particular point, only chlorine gas is available in water. This point is called break point of chlorination.

$$Cl_2 + H_2O \longrightarrow HOCl + HCl$$

$$HOCl + NH_3 \longrightarrow \underset{\text{chloramine}}{NH_2Cl} + H_2O$$

$$HOCl + NH_2Cl \longrightarrow \underset{\text{dichloramine}}{NHCl_2} + H_2O$$

$$HOCl + NHCl_2 \longrightarrow \underset{\text{trichloramine}}{NCl_3} + H_2O$$

1.28 APPLICATIONS OR VARIOUS USES OF WATER

Industry	*Nature of Water*	*Problems*
1. Textile and Dyeing	Free from Fe, Mn, Ca and Mg	It affects the dyeing process
2. Sugar	Free from Sulphates and Alkali carbonates	Crystalisation of Sugar is affected
3. Paper and Pulp	Free from Fe, Mn and SiO_2	Affect Colour, Ash and Cracks in Paper
4. Boiler Feed	Free from hardness and dissolved gases	It causes Boiler Corrosion and Scale
5. Power Station	Free from dissolved gases and scaling agents	It causes Boiler cracks and scale act as insulator

Applications of Engineering in Water Technology

a. Computer software and information technology
   1. Water is used for drinking purpose
   2. So it should satisfy some parameters
   3. Parameters are measured by instruments
   4. Instruments are operated by softwares
      Example, Digital pH Meter, Ti

b. *ECE:* Electronics and Communication Engineering
   1. Instruments are used to characterize water
   2. Instruments are designed by Electronics and Communication Engineering
   3. Voltage conversion is the basic principle of ECE

c. Mechanical
   1. High level or bulk level purification machines are designed by mechanical engineers

d. Electrical Electronics Engineering:
   1. The current convention is an important role.
      So every engineering department acts in an important role on water treatment.

GLOSSARY

Brackish water: Water containing high concentration of dissolved solids.

Coagulant: A substance, which facilitates the coalescence of finely suspended particles.

Disinfection: Destruction of disease producing microorganism to make water safe for use.

Hardness: Soap consuming capacity of water.

Internal treatment: Treatment of water inside the boiler with appropriate chemicals.

Semi permeable membrane: One which permits only solvent.

Sterilization of water: Complete destruction of living organism of water.

Zeolite: A hydrated mixed sodium alumina silicate.

Scale: Hard and adhering precipitate formed in boilers.

Sludge: The soft, slimy and loose precipitate formed inside the boilers.

Molecular Weight of Hardness and Alkalinity Producing Salts:

Molecular Formula of the Salt	Molecular Weight of the Salt
$Ca(HCO_3)_2$	162
$Mg(HCO_3)_2$	146
$Mg(NO_3)_2$	148
$Ca(NO_3)_2$	164
$CaCO_3$	100
$CaSO_4$	136
$MgSO_4$	120
$MgCO_3$	84
$MgCl_2$	95
$CaCl_2$	111
Ca^{2+}	40

Molecular Formula of the Salt	Molecular Weight of the Salt
Mg^{2+}	24
Cl	35.5
C	12
S	32
O	16
H	1
N_2	14
Na	23
NaCl	NaCl does not contribute any hardness to water. Hence it is ignored

Problems based on EDTA method

1. 100 ml of a water sample requires 12 ml of EDTA solution for titration. 1 ml of EDTA Solution is equivalent to 1.1 mgs of $CaCO_3$. Calculate hardness in ppm.

Solution:

$$\begin{aligned}
\text{1 ml of EDTA Solution} &= 1.1 \text{ mgs of } CaCO_3 \\
\text{12 ml of EDTA Solution} &= 12 \times 1.1 \text{ mgs of } CaCO_3 \\
&= 13.2 \text{ mgs of } CaCO_3 \\
\text{100 ml of Water sample required} &= 12 \text{ ml of EDTA} \\
&= 13.2 \text{ mgs of } CaCO_3 \\
\text{1000 ml of Water Sample} &= 13.2 \times \frac{1000}{100} \text{ mgs of } CaCO_3 \\
\text{Hardness} &= 132 \text{ mgs/litre (or) ppm}
\end{aligned}$$

2. 0.3 gm of $CaCO_3$ was dissolved in HCl and the solution was made into one litre with distilled water. 50 ml of the above solution required 30 ml of EDTA solution on titration. 50 ml of hard water sample required 36 ml of same EDTA solution on titration. 50 ml of this water after boiling, cooling and filtering required 12 ml of EDTA solution on titration. Calculate the temporary and permanent hardness of water.

Solution:

$$\begin{aligned}
\text{Given 1000 ml of standard hard water containing} &= 0.3 \text{ gm of } CaCO_3 \\
\text{i.e., 1000 ml of standard hard water contains} &= 0.3 \times 1000 \text{ mg of } CaCO_3 \\
&= 300 \text{ mg of } CaCO_3 \\
\text{1 ml of standard hard water} &= 0.3 \text{ mg of } CaCO_3 \\
\text{30 ml of EDTA} &= 50 \text{ ml of the standard hard water} \\
&= 50 \times 0.3 \text{ mg of } CaCO_3 \\
&= \frac{50 \times 0}{30} \\
\text{1 ml of EDTA} &= 1 \text{ mg of } CaCO_3
\end{aligned}$$

i. Total Hardness:

$$\begin{aligned}
\text{50 ml of hard water} &= 36 \text{ ml of EDTA} \\
&= 36 \times 1 \text{ mg of } CaCO_3 \\
\text{1000 ml of hard water} &= 36 \times \frac{1000}{50} \\
\text{Total hardness} &= 720 \text{ mg / litre or ppm}
\end{aligned}$$

ii. Permanent Hardness (NCH)

$$\begin{aligned}
\text{50 ml of the same water after boiling cooling and filtering required} &= 12 \text{ml of EDTA} \\
&= 12 \times 1 \text{ mg of } CaCO_3 \\
&= 12 \text{ mg of } CaCO_3 \\
\text{1000 ml of water} &= 12 \times \frac{1000}{50} \\
\text{Permanent hardness} &= 240 \text{ mg / litre or ppm}
\end{aligned}$$

iii. Temporary Hardness: (OH)

Temporary hardness = Total hardness – permanent hardness

= 720 – 240

Temporary hardness = 480 mg/litre or ppm

3. 100 ml of a sample of water required 25 ml of an EDTA solution for titration. 30 ml of the same EDTA solution was required for the titration of 100 ml of standard hard water containing 1 gm of $CaCO_3$ per litre. Calculate hardness of water sample in ppm.

Solution:

Given 1 litre of standard hard water contains 1 gm of $CaCO_3$

i.e., 1000 ml of standard hard water contains 1000 mg of $CaCO_3$

1 ml of standard hardness = 1 mg of $CaCO_3$

30 ml of EDTA = 100 ml of standard hard water

= 100 × 1 mg of $CaCO_3$

1 ml of EDTA = $\frac{1000}{50}$ mg of $CaCO_3$

100 ml of sample of water = 25 ml of EDTA

= $25 \times \frac{100}{30}$ mg of $CaCO_3$

for 1000 ml of sample of water = $25 \times \frac{100}{30} \times \frac{1000}{100}$

Hardness = 833.33 mg/lit or ppm

4. Calculate permanent hardness from the following 500 ml of a water sample is boiled for 1 hour. It is then cooled and filtered. The filterate is made upto 500 ml again with distilled water. 50 ml of this solution requires 20 ml of N/50 EDTA with EBT – indicator and NH_4Cl – NH_4OH buffer.

Solution:

Given 50 ml of water sample after boiling filtering requires 20 ml of N/50 EDTA

We know that

1 ml of N/50 EDTA = 1 mg of $CaCO_3$ equivalent hardness

20 ml of N/50 EDTA = 20 mg of $CaCO_3$

50 ml of the boiled water sample requires

= 20 ml of N/50 EDTA

= 20 mg of $CaCO_3$

for 1000 ml of sample of water = $20 \times \frac{1000}{50}$

Permanent hardness = 400 mg/litre of ppm

Problems based on Hardness

1. A sample of water contains 150 mg of $MgSO_4$ per litre. Calculate the hardness interms of $CaCO_3$ equivalents.

Solution:

Given

The amount of $MgSO_4$ = 150 mg/litre

$$\text{Amount equivalent of } CaCO_3 = \frac{\text{The amount of hardness producing salt } x}{\text{molecular weight of hardness producing s}}$$

We know that,

The molecular weight of $MgSO_4$ = 120

$$\text{amount equivalent to } CaCo_3 = 150 \times \frac{100}{120}$$

$$= 125 \text{ mg/litre}$$

2. A water sample contains 272 mg of $CaSO_4$ and 73 mg of $Mg(HCO_3)_2$ per litre. What is the total hardness in terms of CaCO3 equivalnet?

Solution:

$$\text{amount of } CaSO_4 \text{ interms of } CaCO_3 \text{ equivalent} = 272 \times \frac{100}{136}$$

$$= 200 \text{ mg/litre}$$

$$\text{amount of } Mg(HCO_3)_2 \text{ interms of } CaCO_3 \text{ equivalent} = 73 \times \frac{100}{146}$$

$$= 50 \text{ mg/litre}$$

Temporary hardness = $Mg(HCO_3)_2$ = 50 mg/litre

Permanent hardness = $CaSO_4$ = 200 mg/litre

Total hardness = Temporary hardness + Permanent hardness

= 50 + 200

= 250 mg/litre

3. Calculate the carbonate and non - carbonate hardness of a sample of water containing the dissolved salts as given below in mg / litre. $Mg(HCO_3)_2$ = 7.6; $Ca(HCO_3)_2$ = 36.4; $CaSO_4$ = 27.2; $MgCl_2$ = 34.75 and NaCl = 50.

Solution:

$$\left.\begin{array}{r}\text{amount of hardness of } Mg(HCO_3)_2 \text{ in} \\ \text{terms of } CaCO_3 \text{ equivalent}\end{array}\right\} = 7.6 \times \frac{100}{146}$$

$$= 5.20 \text{ mg/litre}$$

$$\left.\begin{array}{r}\text{amount of hardness of } Ca(HCO_3)_2 \text{ in} \\ \text{terms of } CaCO_3 \text{ equivalent}\end{array}\right\} = 36.4 \times \frac{100}{162}$$

$$= 22.46 \text{ mg/litre}$$

$$\left.\begin{array}{r}\text{amount of hardness of } CaSO_4 \text{ in}\\ \text{terms of } CaCO_3 \text{ equivalent}\end{array}\right\} = 27.2 \times \frac{100}{136}$$

$$= 20 \text{ mg/litre}$$

$$\left.\begin{array}{r}\text{amount of hardness of } MgCl_2 \text{ in}\\ \text{terms of } CaCO_3 \text{ equivalent}\end{array}\right\} = 34.75 \times \frac{100}{95}$$

$$= 36.6 \text{ mg / litre}$$

Carbonate hardness

$= Mg(HCO_3)_2 + Ca(HCO_3)_2$
$= 5.20 + 22.46$
$= 27.7$ mg/litre

Non-carbonate hardness

$= CaSO_4 + MgCl_2$
$= 20 + 36.6$
$= 56.6$ mg/litre

Total hardness = Carbonate hardness + Non-carbonate hardness
$= 27.7 + 56.6 = 84.3$ mg/litre

NaCl does not contribute any hardness to water, hence it is ignored.

4. A sample of water contains the following dissolved salts in mg/litre. $Mg(HCO_3)_2$ = 146; $CaCl_2$ = 222; $Ca(HCO_3)_2$ = 100 and $MgSO_4$ = 60. Calculate the temporary and permanent hardness of the water.

Solution:

$$\left.\begin{array}{r}\text{amount of hardness of } Mg(HCO_3)_2 \text{ in}\\ \text{terms of } CaCO_3 \text{ equivalent}\end{array}\right\} = 146 \times \frac{100}{146}$$

$$= 100 \text{ mg/litre}$$

$$\left.\begin{array}{r}\text{amount of hardness of } CaCl_2 \text{ in}\\ \text{terms of } CaCO_3 \text{ equivalent}\end{array}\right\} = 222 \times \frac{100}{111}$$

$$= 200 \text{ mg/litre}$$

$$\left.\begin{array}{r}\text{amount of hardness of } Ca(HCO_3)_2 \text{ in}\\ \text{terms of } CaCO_3 \text{ equivalent}\end{array}\right\} = 100 \times \frac{100}{162}$$

$$= 61.72 \text{ mg/litre}$$

$$\left.\begin{array}{r}\text{amount of hardness of } MgSO_4 \text{ in}\\ \text{terms of } CaCO_3 \text{ equivalent}\end{array}\right\} = 60 \times \frac{100}{120}$$

$$= 50 \text{ mg/litre}$$

Temporary hardness $= Mg(HCO_3)_2 + Ca(HCO_3)_2$
$= 100 + 61.72$
$= 161.72$ mg/litre

Permanent hardness $= CaCl_2 + MgSO_4$
$= 200 + 50$
$= 250$ mg/litre

Problems based on Alkalinity

1. 100 ml of a water sample, on titration with N/50 H_2SO_4 gave a titre value of 7.6 ml to phenolphthalein end – point and 15.2 ml to methyl orange end point. Calculate the alkalinity of the water sample interms of $CaCO_3$ equivalent and comment on the type of alkalinity present.

Solution:

Given

Volume of the water sample = 100 ml
Volume of N/50 H_2SO_4 required to phenolphthalein end point = 7.6 ml
Volume of N/50 H_2SO_4 required to methyl orange end point = 15.2 ml

Since, $P = \frac{1}{2}M$,

The alkalinity of water sample in due to CO_3^{2-} only and not due to OH^- and HCO_3^-

Alkalinity due to CO_3^{2-}

The volume of N/50 H_2SO_4 equivalent to CO_3^{2-} present in 100 ml of the water sample is 2P

$= 2 \times 7.6$ ml $= 15.2$ ml

Volume of the acid consumed V_1 = 15.2 ml
Nomality of the acid N_1 = N/50
Volume of the water sample V_2 = 100 ml
Normality of the water sample due to CO_3^{2-} N_2 = ?

According to volumetric law

$$V_1N_1 = V_2N_2$$

$$N_2 = \frac{15.2}{100} \times \frac{1}{50} = 0.0030$$

Normality of CO_3^{2-} N_2 = 0.00304

Amount of the CO_3^{2-} interms of $CaCO_3$ equivalent $= N_2 \times$ Eq.wt of $CaCO_3 \times 1000$ ppm

$= 0.00304 \times 50 \times 1000$ ppm

Alkalinity of the water sample due to CO_3^{2-} is 152 ppm

2. 200 ml of a sample of water required 20 ml of N/50 H_2SO_4 using methyl orange as indicator but did not give any colouration with phenolphthalein. What type of alkalinity is present? Express the same in mg/litre of $CaCO_3$.

Solution:

Given,

Volume of water sample = 200 ml

volume of N/50 H_2SO_4 required to methyl orange end point } = M = 20 ml

Since $P = 0$, the alkalinity of the water sample is only due to HCO_3^- and not due to CO_3^{2-} and OH^-.

Alkalinity due to HCO_3^-

Volume of the acid consumed V_1 = 20 ml

Normality of the acid N_1 = N/50

Volume of the water sample V_2 = 20 ml

Normality of the water sample due to $HCO_3^{2-}N_2$ = ?

According to volumetric law

$$V_1N_1 = V_2N_2$$

$$N_2 = \frac{20}{200} \times \frac{1}{50} = 0.0030$$

Normality of the HCO_3^- = 0.002

Amount of HCO_3^- interms of $CaCO_3$ equivalent is

= N_2 × Equivalent weight of $CaCO_3$ × 1000 ppm

= 0.002 × 50 × 1000 ppm = 100 ppm

= 100 mg/lit

Alkalinity of the raw water sample due to HCO_3^- = 100 mg/litre

3. 100 ml of a raw water sample on titration with N/50 H_2SO_4 required 13.6 ml of the acid to phenolphthalein end point and 16.2 ml of the acid to methyl orange end point. Determine the type and extent of alkalinity present in the water sample.

Solution:

Given,

Volume of raw water sample = 100 ml

volume of N/50 H_2SO_4 required to phenolphthalein end point } = P = 13.6 m

volume of N/50 H_2SO_4 required to methyl orange end point } = M = 16.2 m

Since $P > ½ M$. the water sample must contain only OH^- and CO_3^{2-} alkalinities and there cannot be any HCO_3^- alkalinity.

i. Alkainity due to OH^-

$$\left.\begin{array}{r}\text{volume of N/50 } H_2SO_4 \text{ equivalent}\\ \text{to OH present in 100 ml of the raw water sample}\end{array}\right\} = 2P - 1$$

$$= (2 \times 13.6) - 16.2 \text{ ml}$$
$$= 27.2 - 16.2 \text{ ml}$$
$$= 11 \text{ ml}$$

Volume of the acid consumed V_1 = 11 ml

Normality of the acid N_1 = N/50

Volume of the raw water sample V_2 = 100 ml

Normality of the raw water sample N_2 = ?

According to volumetric analysis

$$V_1 N_1 = V_2 N_2$$

Substituting the above values, it becomes

$$N_2 = \frac{11}{100} \times \frac{1}{50} = 0.002$$

$$\text{Normality of } OH^- N_2 = 0.0022$$

$$\left.\begin{array}{r}\text{amount of hardness of } OH^- \text{ in}\\ \text{terms of } CaCO_3 \text{ equivalent}\end{array}\right\} = N_2 = \text{equivalent weight of } CaCO_3$$

$$= 0.0022 \times 50 \text{ g}$$
$$= 0.0022 \times 50 \times 1000 \text{ mg}$$
$$= 110 \text{ mg}$$

$$\left.\begin{array}{r}\text{alkalinity of the raw water}\\ \text{sample due to } OH^-\end{array}\right\} = 110 \text{ mg (or) } 110 \text{ P}$$

Alkalinity due to CO_3^{2-}

$$\left.\begin{array}{r}\text{volume of N/50 } H_2SO_4 \text{ equivalent to } CO_3^{2-}OH^-\\ \text{present in 100 ml of the raw water sample}\end{array}\right\} = 2[M - P]$$

$$= 2[16.2 - 13.6] \text{ ml}$$
$$= 2 \times 2.6 = 5.2 \text{ ml}$$

Volume of the acid consumed V_1 = 5.2 ml

Normality of the acid N_1 = N/50

Volume of the raw water sample V_2 = 100 ml

Normality of the raw water sample due to N_2 = ?

According to the volumetric law

$$V_1 N_1 = V_2 N_2$$

Substituting the above values it becomes

$$N_2 = \frac{5.2}{100} \times \frac{1}{50} = 0.0010$$

$$\text{Normality of } CO_3^{2-}\ N_2 = 0.00104$$

$$\left.\begin{array}{r}\text{amount of } CO_3^{2-} \text{ in}\\ \text{terms of } CaCO_3 \text{ equivalent}\end{array}\right\} = N_2 = \text{equivalent weight}$$

$$\text{of } CaCO_3 \times 1000 \text{ pp}$$

$$= 0.00140 \times 50 \times 1000 \text{ ppm}$$

$$= 52 \text{ ppm}$$

ii. Total Alkalinity

$$\text{Total alkalinity} = \text{alkalinity due to } OH^- + \text{alkalinity due to } CO_3^{2-}$$

$$= 110 \text{ ppm} + 52 \text{ ppm}$$

$$= 162 \text{ ppm}$$

EXERCISES

Problems Based on Hardness

1. A sample of water contains 60 mg of $MgSO_4$ per litre. Calculate the hardness in terms of $CaCO_3$ equivalent.

   [Ans: 50 mg/litrre]

2. If a sample of water contains 80 mg of Ca^{2+} ions per litre, calculate its hardness in terms of $CaCO_3$ equivalent.

   [Ans: 200 mg/litre]

3. A water sample contains 204 mg of $CaSO_4$ and 73 mg of $Mg(HCO_3)_2$ per litre. What is the total hardness in terms of $CaCO_3$ equivalent?

   [Ans: 200 mg / litre]

4. A sample of water is found to contain the following analytica data in mg/litre.

$$Mg(HCO_3)_2 = 14.6$$
$$MgCl_2 = 9.5$$
$$MgSO_4 = 6.0$$
$$Ca(HCO_3)_2 = 16.2$$

5. Calculate the temporary and permanent hardness of the sample of water. (Atomic weight: Ca = 40, Mg = 24; Cl = 35.5; C = 12; S = 32; O = 16 and H = 1)

   [Ans: 15 mg/litre]

Problems based on Alkalinity

1. 100 ml of a raw water sample, on titration with 0.02 N sulphuric acid required 10 ml of the acid to phenolphthalein end point and 16.0 ml of the acid to methyl orange end point. Determine the type and the extent of alkalinity present in the water sample.

   [Ans: 160 ppm]

2. 200 ml of a water sample required 12 ml of 0.02 N H_2SO_4 using phenolphthalein indicator but it directly gives pink colouration with methyl orange. What type of alkalinity is present?

[Ans: 60 ppm]

3. 100 ml of a water sample, on titration with N/50 H_2SO_4 gave a titre value of 6.8 ml of phenolphthalein end point and 13.6 ml to methyl orange end point. Calculate the alkalinity of the water sample in terms of $CaCO_3$ equivalent and comment on the type of alkalinity present.

[Ans: 136 ppm]

Problems based on EDTA Method

1. 100 ml of a water sample required 30 ml of EDTA solution for titration (1 ml of EDTA = 1 mg of $CaCO_3$). Calculate the hardness of the sample of water.

[Ans: 300 ppm]

2. 100 ml of a sample of water required 25 ml of 0.01 M EDTA for the titration using Eriochrome - Black - T indicator. Calculate the total hardness.

[Ans: 250 ppm]

3. 100 ml of water sample required 20 ml of 0.01 M EDTA for the titration with Eriochrome Black-indicator. 100 ml of the same sample after boiling and filtering required 10 ml of 0.01 M EDTA. Calculate the total carbonate and non-carbonate hardness of the sample.

[Ans: 100 ppm of $CaCO_3$ equivalent]

4. In an estimation of hardness of water by EDTA titration, 250 ml of a sample of water required 15 ml of 0.025 M EDTA solution to get to the end point. Calculate the hardness of water.

[Ans: 150 ppm]

5. 25 ml of standard hard water containing 1 mg of $CaCO_3$ per ml consumes 12 ml of standard EDTA Solution. 25 ml of hard water consumes 8 ml of standard EDTA solution. After boiling the sample, 25 ml of the boiled hard water consumes 6 ml of standard EDTA solution. Calculate the total, permanent and temporary hardness.
6. [Ans: Total Hardness = 666.67 ppm, Permanent Hardness = 500 ppm, Temporary Hardness = 166.67 ppm]

ASSESSMENT QUESTIONS

Questions (2 Marks)

1. Define hardness of water.
2. What are the salts responsible for carbonate and non-carbonate hardness of water?
3. What are the constituents that cause hardness in water?
4. Explain the terms "carbonate hardness" and "non-carbonate" hardness
5. Distinguish between hard and soft water.

6. Distinguish between carbonate hardness (ch) and non-carbonate hardness (nch)
7. How is hardness of water expressed?
8. Bring out the significance of calcium carbonate equivalents?
9. A water sample contains 73 mg $(HCO_3)_2$ per litre. Calculate the hardness in terms of $CaCO_3$.
10. Draw the structure of EDTA. what happens when EDTA is added to hard water?
11. How does Eriochrome-Black-T indicator function as an indicator in EDTA titration?
12. Why NH_4Cl-NH_4OH buffer is used in EDTA titration.
13. What is meant by permanent hardness of water? Mention the salts responsible for the permanent hardness of water.
14. Define alkalinity.
15. How is alkalinity classified?
16. How is alkalinity determined?
17. Why is water softened before using in boiler?
18. What are scales and sludges?
19. What is meant by priming and foaming? How can they be prevented?
20. What is meant by caustic embitterment? How is it prevented?
21. indicate the reasons for boiler corrosion
22. What are the requisites of drinking and boiler feed water.
23. Define the advantage of softening water. How is it carried out?
24. soft water is not D M water whereas DM water is soft water-justify
25. What are the advantages of ion exchange process?
26. How is exhausted resin regenerated in the ion exchange process?
27. Give some examples for cation exchange resin.
28. Give some examples for anion exchange resin.
29. How is water demineralised in an ion-exchanger?
30. Give the description of sand filter.
31. What is aeration of water? mention its purpose
32. Explain the functions of a coagulant with examples.
33. What is carbonate conditioning? For what type of boilers is it used?
34. What is the role of phosphates in the internal treatment of water?
35. What are boiler compounds? Mention two different boiler compounds and their actions.
36. What are calgon conditioning? How is it functioning in water treatment?
37. Mention requisites of potable water.
38. What are the various stages in the treatment of water for domestic supply?
39. Write briefly on disinfection of water by UV treatment
40. Explain the function of bleaching powder as a germicide
41. Mention the requisites of water used as potable water
42. What is meant by "reverse osmosis"? How is it applied in the desalination of water?

43. Define the term breaking point chlorination.
44. Name some membranes employed in the reverse osmosis process. State the advantages of the process.
45. What are the advantages of the reverse osmosis method?
46. What is blow-down operation?
47. what are the disadvantages of ion exchange process?
48. Define desalination.
49. What are the disadvantages of using ozone in disinfection of water?
50. Name the method which separates both ionic and non-ionic impurities from water

Questions (16 Marks)

1. What is meant by carbonate and non-carbonate hardness of water? Explain with examples.
2. What is the principle of EDTA method? Describe the estimation of hardness of water by the EDTA method.
3. Describe the principle and method involved in the determination of different types and amount of alkalinity of water.
4. Describe briefly the different steps in the purification of water for drinking purposes. What is the usage of breakpoint chlorination?
5. Define the term desalination with a neat diagram, describe desalination by the reverse osmosis method
6. Describe de-mineralization process of water softening and explain the reactions involved
7. How is internal treatment of boiler water carried out using phosphate and calgon?
8. Discuss the causes and prevention of priming and foaming.
9. What is the principle of reverse osmosis? How is it used for desalination?
10. What are scales and sludges? Describe the disadvantages of scale and sludge formation
11. What are the various methods by which disinfection of domestic water is carried out and explain?
12. Write a note on desalination by reverse osmosis and internal treatment of boiler water.
13. Explain phosphate and carbonate methods of internal treatment of boiler water.
14. How will you protect a boiler from corrosion.
15. Discuss in detail the disadvantages of using hard water in boilers.
16. Describe the internal conditioning of water.
17. Explain in detail the demineralization process for softening of water.
18. What is meant by reverse osmosis? Explain purification of water by reverse osmosis.
19. What is desalination? Name the different methods of desalination. Explain any one in detail.

20. Explain the determinations of various alkalinity of water sample.
21. Write notes on
    Carbonate and phosphate conditioning
    Types of hardness
22. What is boiler feed water? Explain the various boiler troubles.
23. Explain the following:
    i. Estimation of hardness by EDTA method
    ii. Reverse osmosis
24. What is potable water? What are the steps taken to obtain pure drinking water?
25. Explain boiler troubles? How are they caused? Suggest steps to minimize boiler troubles.

2

Polymer and Composites

2.1 INTRODUCTION

About the polymer we can describe by the term the land of neglected dimensions to describe the range of sizes between molecular and macroscopic within which occur most colloidal particles.

The term neglected dimension refers to a polymer molecule. It was not since 1930.

We can polymers classify into two types:

1. Biological polymer
2. Non-biological polymer

Biological polymers provide food for human beings. Non-biological polymers are artificial polymers.

Based on the polymer availability, polymers are classified into two types:

Nowadays based on the Polymers, technology and engineering materials are developing.

For an example Medical instruments, Computer, Vehicles spare parts, Furniture and Air Crafts.

Flow chart 2.1

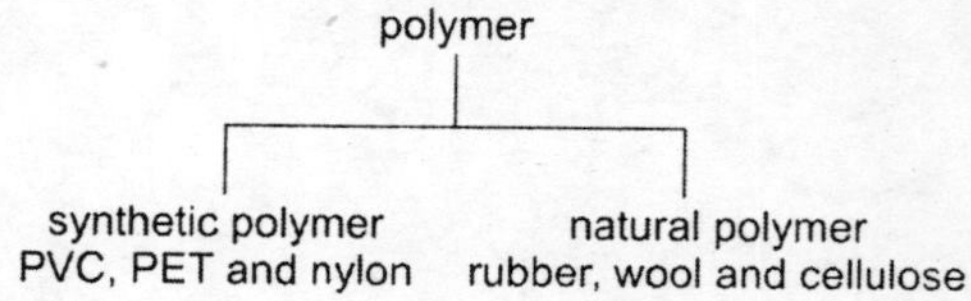

2.2 DEFINITION FOR POLYMER

Polymer is explained by different ways. Poly means "many" and mers mean "units". Monomer units merge together to form a polymer. More than one monomer units are linked by chemical reactions. The product is called a polymer. Materials which have more than one repeatable units is called polymer.

Definition for Polymerisation

The process of Making or preparing a polymer is called polymerization. More than one monomer units are linked by chemical reaction process called polymerization. Making a very big molecule by linking monomer units is called polymerization.

2.3 MONOMER

Monomer is a reverse of polymer. The repeatable unit of a polymer is called a monomer.

Based on the functional groups, monomers are classified into saturated monomers and unsaturated monomers.

Saturated Monomer

These monomers form polymers by their active functional groups. It has two functional groups without conjugation or a double bond.

Example,

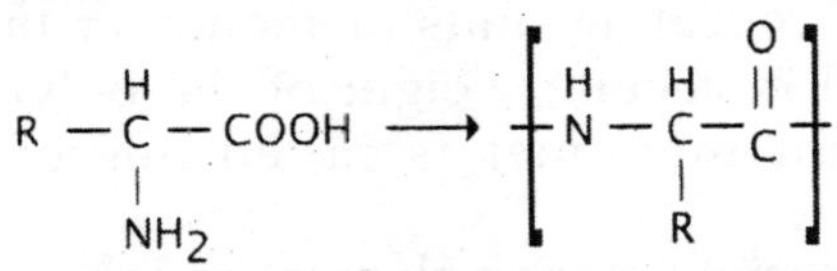

amino acid → polypeptide

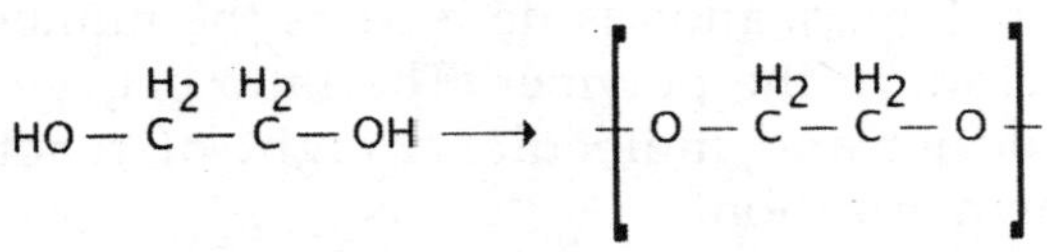

ethylene glycol → poly ethylene glycol

Unsaturated Monomer

This monomer has Alkenes or double bonds in it. This double bond is the active functional group of the monomer.

Example, *Monomer* *Polymer*

1.

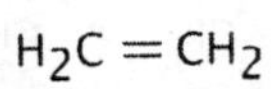

ethylene

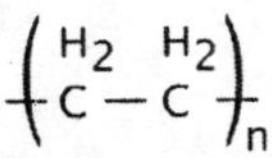

poly ethylene

2.

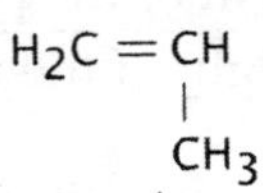

propylene

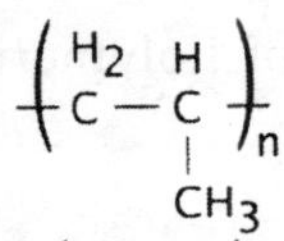

poly propylene

3.

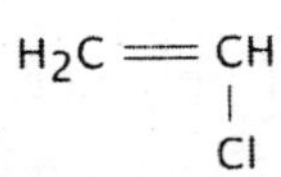

vinyl chloride

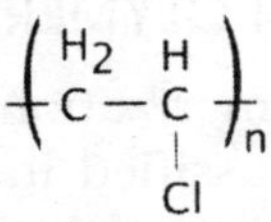

polyvinyl chloride

4.

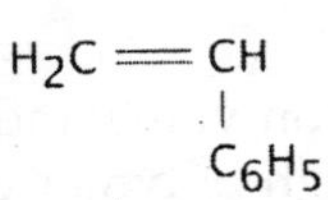

styrene

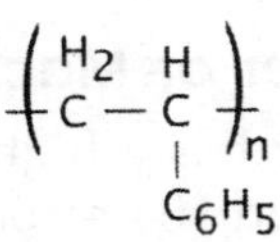

poly styrene

5. 

tetra fluoro ethylene — poly tetra fluoro ethylene

6. 

vinyl alcohol — poly vinyl alcohol

2.4 DEGREE OF POLYMERISATION

The polymer product nature is calculated by the degree of polymerization. Degree of polymerization is defined as the number of repeating units or monomer units present in the polymer. The ratio between the molecular weight of the polymer product and molecular weight of reactant or monomer is called degree of polymerization.

$$\text{Degree of Polymerisation} = \frac{\text{molecular weight of the polymer product}}{\text{molecular weight of the monomer}}$$

$$= \frac{\text{molecular weight of the polymer product}}{\text{molecular weight of the repeatable unit}}$$

Example

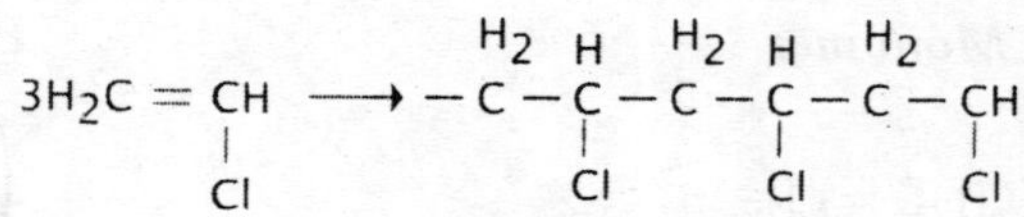

monomer mw - 72 — polymer product mw - 216

$$\text{Degree of Polymerisation} = \frac{\text{molecular weight of the polymer product}}{\text{molecular weight of the repeatable unit}}$$

$$= \frac{216}{72} = 3$$

2.5 TYPES OF POLYMERS

Degree of polymerization explains the character of polymer. Based on this value polymers are classified into two types:

1. High Polymers or Macromolecules
2. Lower Polymers or Oligo polymers

High Polymer or Macromolecules

If the polymer has a higher degree of polymerization value and a higher molecular weight between 104 – 106 gm, then it is called a high polymer or a macromolecules

Oligomer or Low Polymer

Lower molecular weight and lower degree of polymerization polymer is called oligomer.

2.6 TYPES OF POLYMERISATION

Based on the chemical reactions by which the polymer is formed can be classified into three types:

1. Addition Polymerisation
2. Copolymerisation
3. Condensation Polymerisation

2.7 ADDITION POLYMERISATION

Making polymers without the formation of by products is called addition polymerisation. From this process multiplicity of the same reactant monomeric unit occurs and there is no loss of a molecule in the reaction so by -products are not formed. Simple bond transformation occurs. This process is called addition polymerization and the polymer product is called addition polymer.

Example,

1. Poly Ethylene

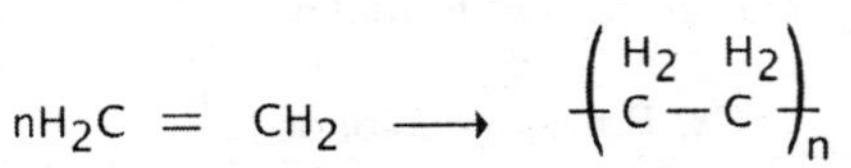

ethylene poly ethylene

2. Poly Styrene

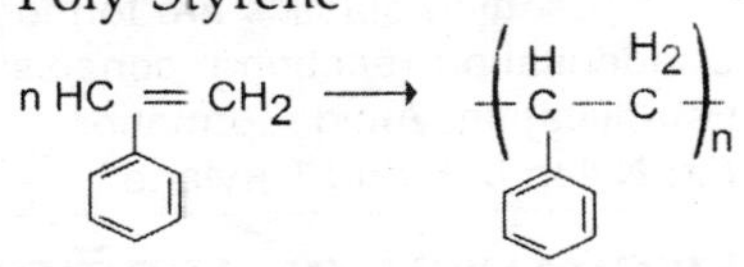

styrene poly styrene

2.8 COPOLYMERISATION

Polymerization process is carried out with two different monomers and without a loss of molecules is called copolymerization. There is an Equal reactivity of different monomers.

Example,

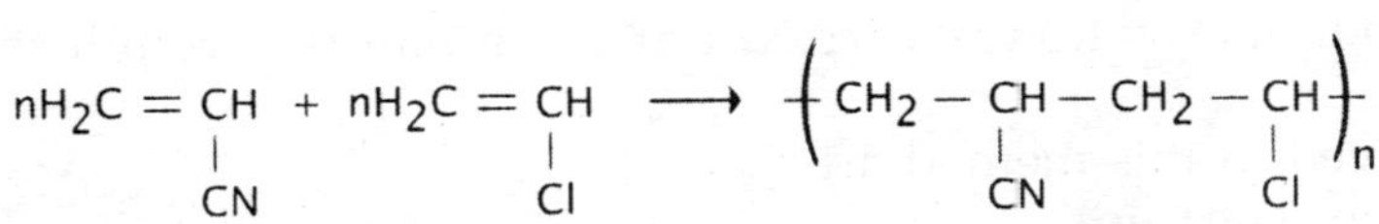

monomer I acrylonitrile monomer II vinyl chloride polymer

2.9 CONDENSATION POLYMERISATION

Polymerization process is carried out with two monomers and two different active centers is called condensation polymerization. During this process, loss of molecular weight from monomers occurs and by-products are condensed. Condensed molecules may be CH_3OH, H_2O etc.For this process heat energy or catalysts are required to forward the reaction.

Example, Terylene, Nylon

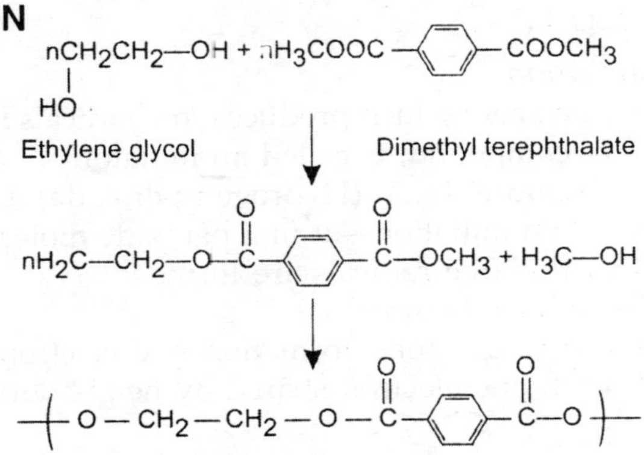

Polyethylene terephthalate

2.10 DIFFERENCES BETWEEN ADDITION AND CONDENSATION POLYMERIZATION

No.	Condensation	Addition
1.	Two different active centres involved	Only one active centre
2.	By-products are condensed	No by products formed
3.	Hetero chain (different monomer molecule) polymer is formed	Homo chain (single monomer) polymer is formed
4.	Two functional groups are present at the end of the molecule ***Ex:*** $HO-\overset{H_2}{C}-\overset{H_2}{C}-OH$	Double bond is the active centre ***Ex:*** $H_2C=CH_2$
5.	Chain growth at both the ends of the monomer occurs	Chain growth at only one centre
6.	Rate of the reaction is slow	Rate of the reaction is fast
7.	Product is formed stepwise	Product is formed fast
8.	Thermosetting plastics are formed	Thermo plastics are formed
9.	Condensation reaction mechanism like Esterification, Amide formation	Free radical formation
10.	***Ex:*** Nylon 6, 6 and Terylene	***Ex:*** Polyvinyl chloride

2.11 MECHANISM OF ADDITION POLYMERIZATION

Type of polymerization: Addition Polymerisation
Mechanism: Free Radical Mechanism
Catalyst: Organic Peroxides

Example, Benzoyl peroxide, Acetyl peroxide, Hydrogen peroxide

Mechanism in three steps

- Addition polymerization is carried out by initiators through the free radical mechanism.
- Steps involved in this method are:
  1. Initiation (start up)
  2. Propagation (growing of polymer)
  3. Termination (end of the reaction)

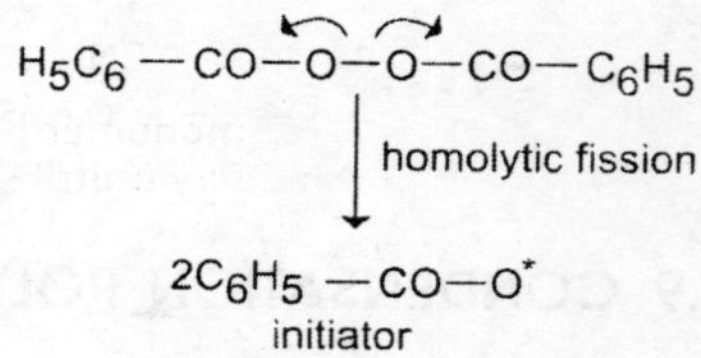

Step 1

Initiation

- The agent which produces free radicals from reactant molecules by decomposition of compound, is called an initiator.
  Example, H_2O_2 (Hydrogen peroxide), $C_6H_5COOOCOC_6H_5$ (benzoyl peroxide)
- Chain initiation—at first peroxide molecule undergoes homolytic fission. Because of this free radicals are formed.

Homolytic Fission

- For Single bond formation two electrons are required.
- Each one electron shared by neighbouring atom in molecule is called homolytic fission 

Heterolytic Fission

Two electrons taken by only one neighbouring atom is called heterolytic fission

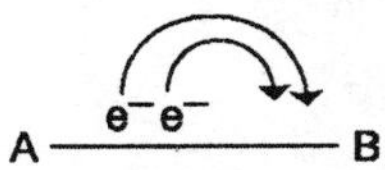

Initiator reacts with the monomer and form a free radical by addition.

$$R^* + H_2C = CH_2 \longrightarrow R - \overset{H_2}{C} - C^*H_2$$

Monomer — first free radical

Initiation Step

Chain Propagation

In this step, the first free radical react with "n" number of monomers.

$$R - CH_2 - \overset{*}{C}H_2 + n\,H_2C = CH_2$$
$$\downarrow$$
$$R - \left[CH_2 - CH_2 \right]_n CH_2 - \overset{*}{C}H_2$$

Chain Termination

Based on the end product, two possibilities exist. Lastly, final product formation is called the termination reaction.

1. Coupling or Combination
2. Disproportionation

Step 2

Coupling or Combination

Same free radical molecules react together and form higher polymers. This is called coupling reaction. The product is called a dead polymer.

$$R - \left[CH_2 - CH_2 \right]_n CH_2 - \overset{*}{C}H_2 + \overset{*}{C}H_2 - CH_2 \left[CH_2 - CH_2 \right]_n - R$$
$$\downarrow$$
$$R - \left[CH_2 - CH_2 \right]_n CH_2 - CH_2 - CH_2 - CH_2 \left[CH_2 - CH_2 \right]_n - R$$

dead polymer

Disproportionation

- Same molecules are involved in the hydrogen transformation reaction.
- By this method two different polymers are formed.
- One is saturated and another one is unsaturated.

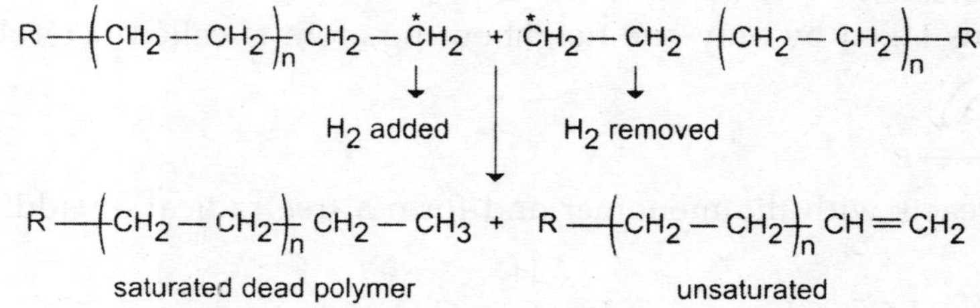

Individual Polymer Preparations, Properties and Uses

2.12 POLY VINYL CHLORIDE (PVC)

Type of Polymerisation: Addition Polymerisation
Mechanism: Free Radical Mechanism
Monomer: vinyl chloride.

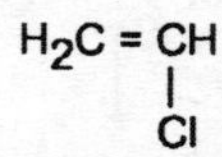

vinyl chloride

Catalyst: Hydrogen peroxide(or) Benzoyl peroxide
Process: Heating of water emulsion of Vinyl chloride in the presence of an initiator in an autoclave at pressure of 1 – 1.5 atm.

Reaction

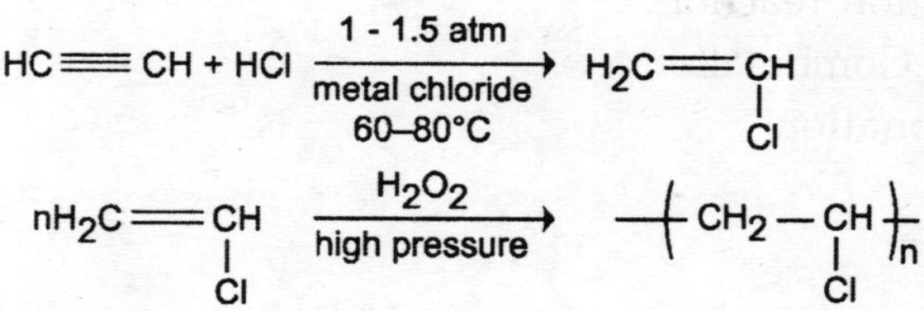

Properties

- Colourless and odourless
- Chemically inert, non-inflammable
- Resistant to light, atmospheric oxygen, in-organic acids, alkalis
- High softening point at 148°C
- Higher stiffness

Types of PVC

Based on the additives added with the polymer, it is classified into two types:

1. Rigid PVC (or) unplasticised PVC.
   - Polyvinyl chloride
2. Plasticized PVC
   - PVC treated or added with Dibutyl phthalate, Tricresyl phosphate, Dioctyl phthalate is called plasticized PVC
   - For making a continuous sheet in production (0.1 mm to 8 mm thickness)

Uses

1. Rigid PVC used to prepare safety helmets, light fittings, tyres etc.
2. Plasticized PVC used to prepare sheets.
3. It is used for making rain coats.
4. Chemical containers and conveyor belts
5. In air crafts, cinemas used as thermal insulating foam

2.13 TEFLON: POLYTETRAFLUOROETHYLENE (PTFE) OR FLUON

Types of polymerization: Addition Polymerisation
Mechanism: Free Radical Mechanism
Monomer: Tetra Fluoro Ethylene

Reaction

i. Hydrogen Fluoride

$$CaF_2 + H_2SO_4 \longrightarrow CaSO_4 + 2HI$$

ii. Chloroform

$$CH_4 + 3Cl_2 \xrightarrow[\text{light}]{h\nu} CHCl_3 + 3HCl$$

iii. Tetra Fluoro Ethylene

$$CHCl_3 + 2HF \xrightarrow{SbF_3} CHClF_2 + 2HCl$$

$$2CHClF_2 \longrightarrow F_2C = CF_2 + 2HCl$$

Properties

- Due to the presence of the highly electronegative atom—fluorine, teflon has good properties.
- High toughness
- High softening point (350°C)
- High chemical resistant
- High density
- Good electrical and mechanical properties

Uses

- It is used as an insulating material
- It is used for making gaskets, pump parts
- Coating the fiber glass and asbestos
- Used as non-lubricating bearings, non-stick frying-pans and stop clocks.

2.14 POLY AMIDES

Amide Functional Group

$$-\overset{\overset{\displaystyle O}{\|}}{C}-NH_2$$

If the repeating monomer unit ends with amide functional groups in polymer then it is called polyamide.

Example, Nylon 6 and Nylon 6:6

Based on the carbon number, in-between amide functional group, polymer ends with the number

Nylon 6

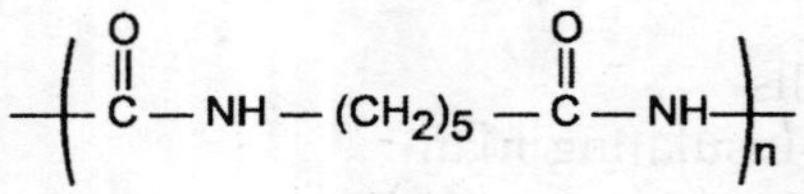

Nylon 6:6

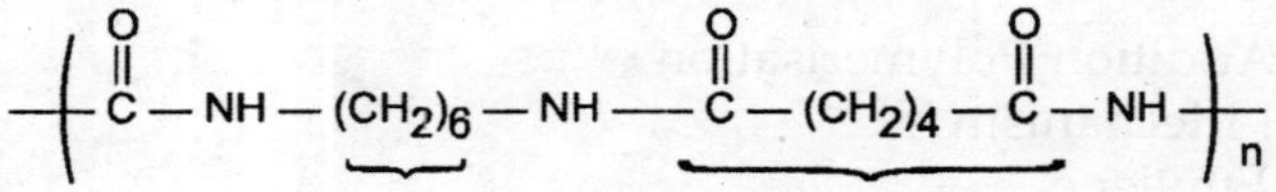

Preparation of Poly Amide

Type of polymerization: Condensation polymerization
Mechanism: Amide Formation
Monomer: Hexamethylene diamine, Adipic acid

Reaction for Nylon 6

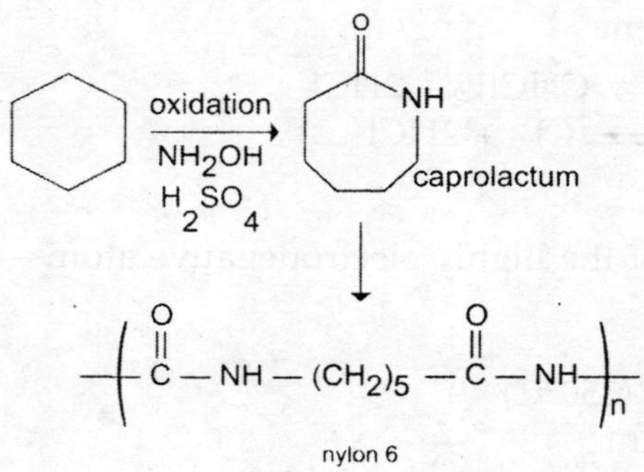

Nylon 6:6

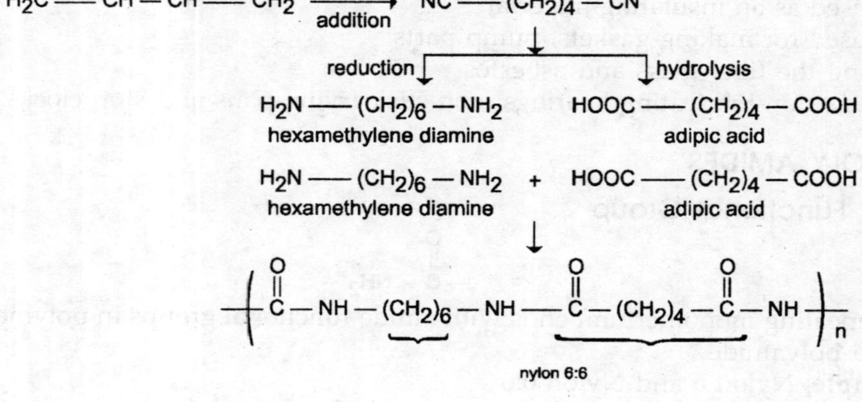

Properties

- It is translucent (Partially allowing light or radiation)
- It is white in colour
- It is horny (hard material) in nature
- Temperature stability
- Good abrasive resistance
- Self lubricating agent
- Good physical strength and insoluble in common organic solvents

Uses

- Making socks, garments, carpets
- Making gears, bearings
- To prepare ropes, tooth brushes

2.15 POLY ETHYLENE TEREPHTHALATE (PET)

Type of polymerisaiton: Condensation of Polymerisation
Monomers: Ethylene glycol, Terephthalic acid
Mechanism: Esterification

Reaction

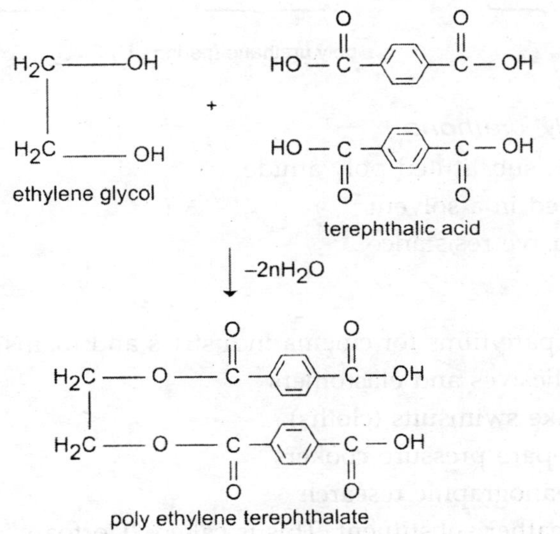

Properties

- It is a good fiber forming material
- It has a high stretch resistance, high crease and wrinkle resistance (Crease - after folding, lines do not appear)
- It is highly resistance to minerals and organic acids but less to alkalis

Uses

- It is used for making synthetic fibers like terylene cloths, dacron etc.

- It is used for blending with wool because of its high crease wrinkle resistance character
- It is used as a glass reinforcing material
- It is used in helmets, Aircrafts and battery boxes, etc.

2.16 POLY URETHANE OR PERLON - U

Type of polymerization: Condensation polymerization
Monomers: Hexa methylene di isocyanate and 1, 4-butane diol
Mechanism: Hydride shifting and Amide formation

Reaction

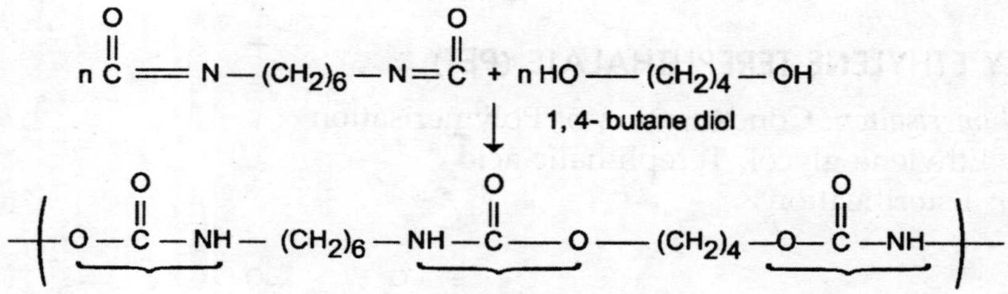

poly urethane (perlon - U)

Properties of Poly urethane

- It is a less stable substituted poly amide
- It is not dissolved in a solvent
- It is highly abrasive resistance

Uses

- It is used to prepare films for cinema industries and foams
- It is used as adhesives and elastomers
- It is used to make swimsuits (cloths)
- It is used to prepare pressure cookers
- It is used in oceanographic research
- It is used as a leather substituent. This is called "Corfoam"

2.17 POLY CARBONATES (PC) OR LEXAN OR MERLON

*Types of polymerization:*Condensation polymersation
Monomers: Bisphenol - A, Phosgene or Diphenyl carbonate, Bisphenol - A
Functional group:

$$RO-\overset{\displaystyle O}{\overset{\|}{C}}-OR$$

Reaction

Type 1

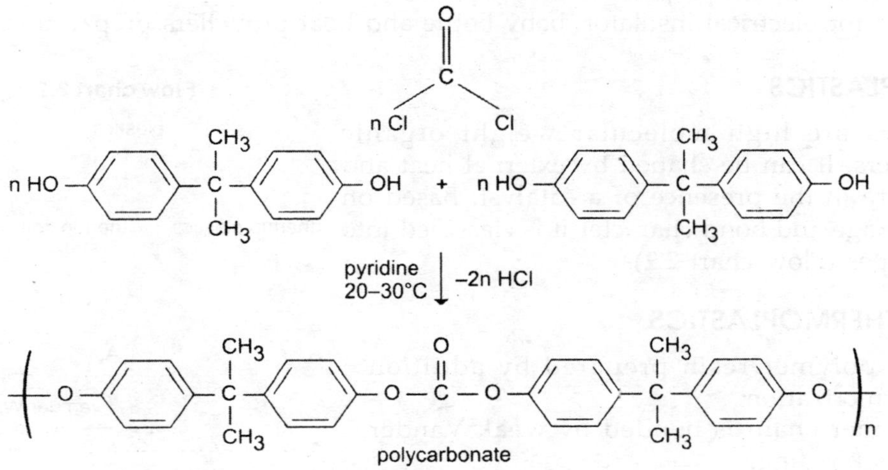

In this reaction HCl is formed. This is neutralized by pyridine.

Type 2

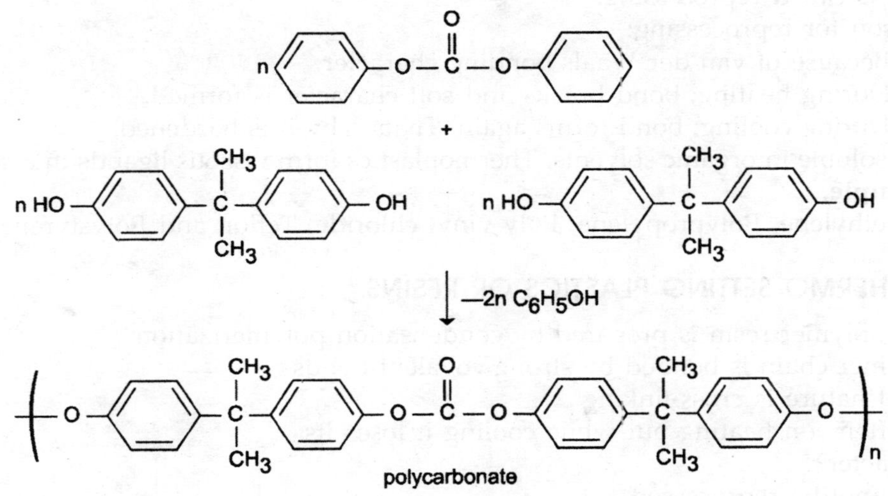

Properties

- Poly carbonates have good mechanical and tensile strength
- It is soluble in organic solvents and alkalies
- It has good stiffness and transparency

Uses

- It is used to prepare safety goggles
- It is used for the preparation of telephone parts
- Used for electrical insulator, baby bottle and boat propellers preparation

2.18 PLASTICS

Plastics are high molecular weight organic polymers. It can be shaped by external heat and pressure in the presence of a catalyst. Based on the linkage and bond character it is classified into two types (Flow chart 2.2).

Flow chart 2.2

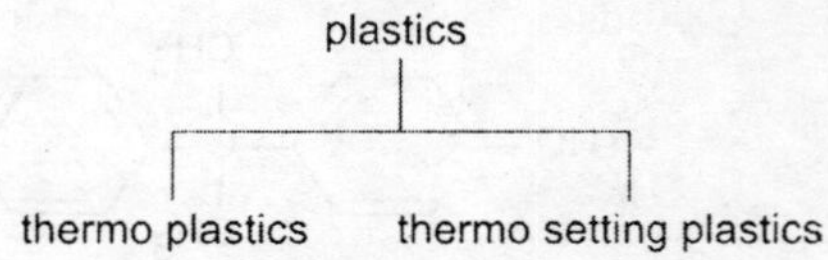

2.19 THERMOPLASTICS

- The polymer resin prepared by addition polymerization.
- Polymer chain is bonded by weak Vander walls bonding.
- Bond nature is linear (or) branched chain nature.
- Low molecular weight polymer.
- It is soften on heating and harden on cooling. This is called reprocessing.
- Reason for reprocessing:
  1. Because of van der Waals bonding character.
  2. During heating, bond breaks and soft character is formed.
  3. During cooling, bond forms again. Thats why it is hardened.
- It is soluble in organic solvents. Thermoplastics form viscous ligands in a solvent.
  Example,
  Polyethylene, Polypropylene, Poly vinyl chloride, Teflon and Polystyrene.

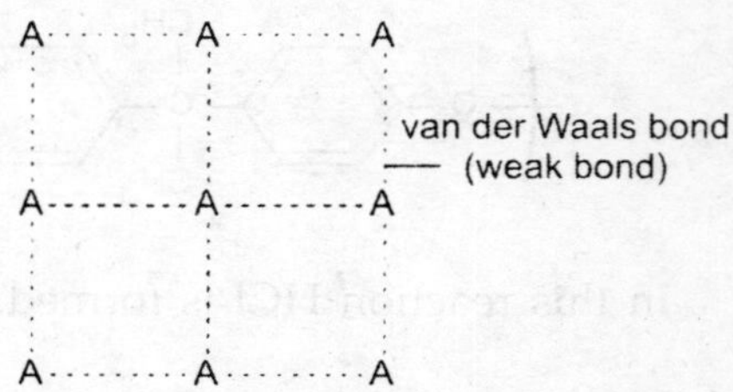

Fig. 2.1

2.20 THERMO SETTING PLASTICS OR RESINS

- The polymer resin is prepared by condensation polymerization
- Polymer chain is bonded by strong covalent bonds
- Bond nature is cross-linking
- It softens on heating but while cooling it loses its character
- It cannot be reprocessed.
- Reason for the above point is that while heating the covalent bond is broken
- It is insoluble in organic solvents
- High molecular weight polymers
  Example,
  i. Phenol - formaldehyde resin

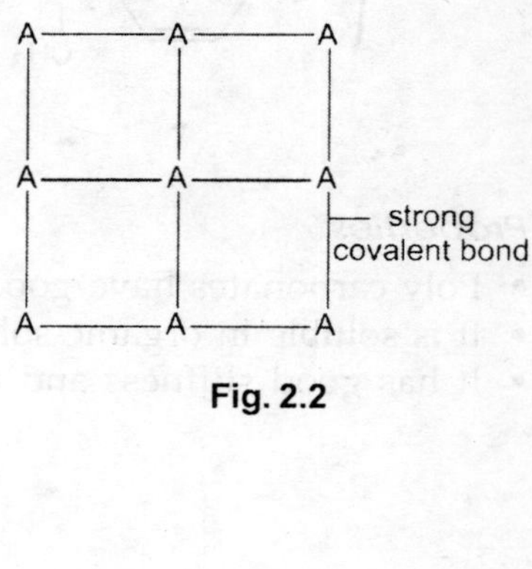

Fig. 2.2

ii. Epoxy resin
iii. Melamin formaldehyde resin
iv. Terylene

Differences between Thermosetting and Thermoplastics:

No.	Thermoplastics	Thermosetting Plastics
1.	Addition polymerization product	Condensation polymerization product
2.	Weak van der Waals bond between neighbouring molecule	Strong covalent bonded molecule
3.	On heating softens and on cooling hardens	It is charring
4.	Remoulding is possible	Remoulding is not possible
5.	They are weak, soft and less brittle (Semi powder nature)	They are strong hard and more brittle
6.	It is soluble in organic solvent	It is insoluble
7.	Weak van der Waals bond form again	Cross link strong covalent bond breaking
8.	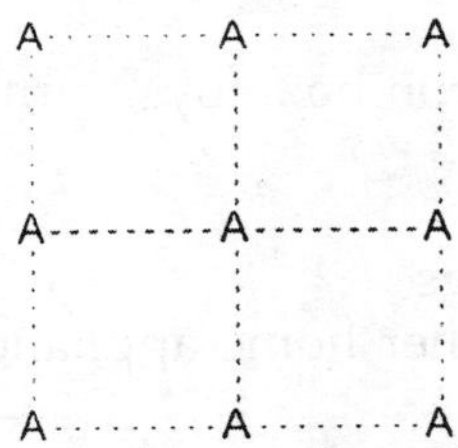	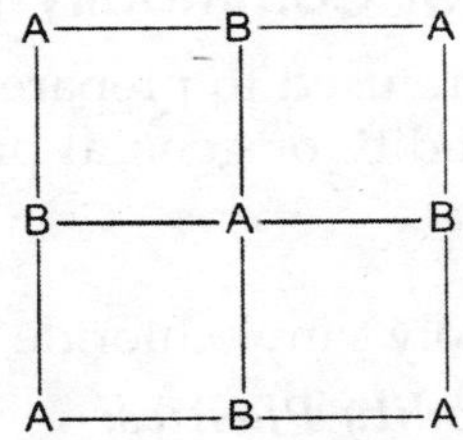

2.21 GLASS TRANSITION TEMPERATURE (TG)

A polymer is a solid material. On heating the polymer state is changed. Glass transition temperature is that at which particular temperature, a solid polymer gets converted to a liquid polymer. Below the glass transition temperature, the polymer is hard and it is called glassy state. Above the Tg it is soft, this state is called as the rubbery (or) visco elastic state. Symbol for glass transition temperature is Tg.

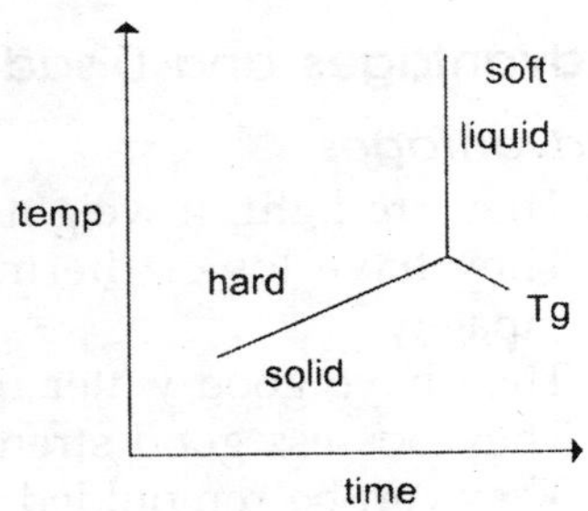

Fig. 2.3

Types of plastics based on application:

i. Engineering plastics
ii. General purpose plastics (or) commodity plastics

2.22 ENGINEERING PLASTICS

1. Engineering plastics are prepared from high polymer resins.
2. For engineering usage purpose it has Mechanical Strength,Roughness and High thermal resistance.
3. Nowadays they are replacing metal, wood, glass and ceramics in the engineering field
   Example, initially, cars were made up of metal ions

4. But nowadays, maximum parts are made by polymer plastics.
5. Polymers are mixed with metals
   Example, Four wheeler radiator circulation tube
6. Because of the above points, it is called Engineering plastics.

Properties of engineering plastics

1. They have high softening (previous stage of melting temperature) temperature.
2. High thermal resistance (above 180°C)
3. They posses good mechanical strength and can be molded into any shape
4. They have load bearing capacity, light weight and a high abrasive resistance
5. They have good crystalline nature and strong bond nature
6. They are not affected by chemicals and weather

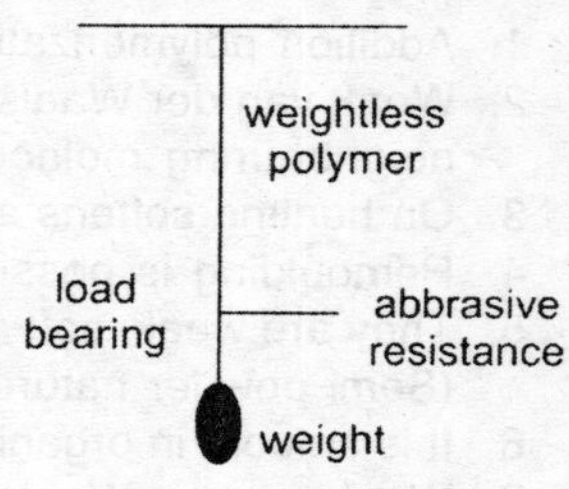

Fig. 2.4

General Purpose or Commodity Plastics

The plastics which are used to prepare home materials like iron box, toys, furniture etc are called commodity or general purpose plastics.

Example Polymers:

Poly ethylene – Covers, Carry bags
Poly propylene, Poly vinyl chloride – Toys and some other home appliances

Properties of Commodity Plastics:

- Low molecular weight polymers are used.
- Lower glass transition temperature (Tg) and lower thermal resistance.
- They have lower mechanical strength.
- They have poor dimensional stability (arrangement or structure changing).

Dimensional stability: Plastic toys change their structures under pressure or by breaking them.

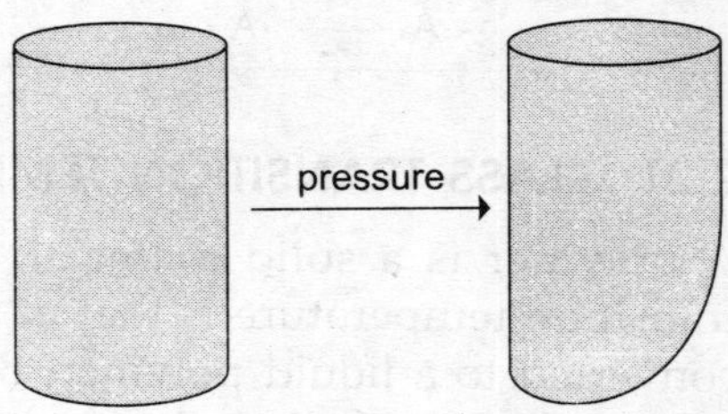

Fig. 2.5

Advantages and Disadvantages of Plastics

Advantages

- They are light in weight and low melting solids
- They have lower thermal expansion co-efficients and good shock absorption capacity
- They have good water resistance and chemical resistance
- They possess good strength and toughness
- They can be remoulded

Disadvantages

- They cause pollution
- They become embrittled at low temperatured (Powder nature)
- Easily combustible and emitting higher amounts of CO_2 and CO smoke
- These are all non biodegradable

2.23 RUBBER

Nowadays rubber material acts an important role, because of their high elasticity. Rubber is non crystalline, high molecular weight polymer. It is a linear molecule. It has a higher elasticity character, that's why it is called elastomer or rubber

Difference between Plastics and Rubber

No.	Plastics	Rubber
1.	It is crystalline nature polymer	It is non crystalline
2.	Elasticity is very less	Elasticity is more

Types of Rubber

Based on the availability, preparation process and source, rubber can be classified into two types:

1. Natural Rubber
2. Artificial or Synthetic Rubber

Natural Rubber

Source: Latex

Plant or Tree: Hevea brasiliensis

Latex: It is an aqueous dispersion of rubber. It contains 25–40% rubber hydrocarbon, a small amount of protein and fatty acids.

Process

Latex is treated with sodium bi sulfite for bleaching.Then the coagulum is rolled out into sheets and dried in air at about 50°C. The liquid contain more isoprene molecule on drying. It polymerises and form polyisoprene long coiled chain.

Reaction

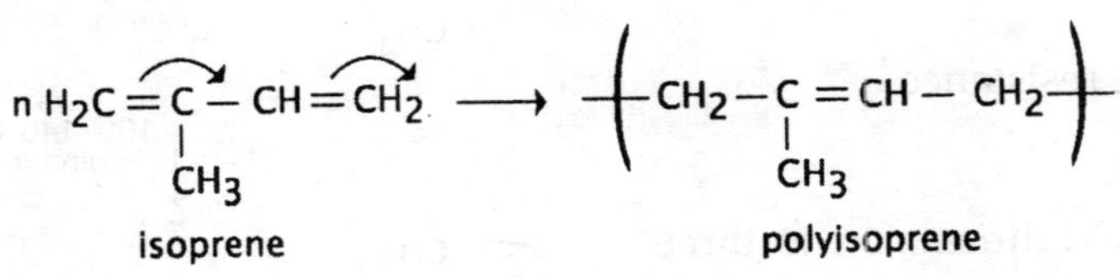

Raw rubber which is obtained from latex has some drawbacks

It is soluble in organic solvents and oils. It has little durability and poor elasticity. It is plastic in nature, poor strength and water absorption character. It is oxidized by Nitric acid (HNO_3) and Sulphuric acid (H_2SO_4)

Quality of Rubber Improved by Following Method

Because of the drawbacks in (latex) raw rubber, its properties and quality must be improved.

2.24 VULCANISATION DEFINITION

The method used to improve the quality of raw rubber by adding sulphur is called vulcanisation.

Vulcanisation Process

Raw rubber is treated with Sulphur (S), Hydrogen sulphide (H_2S) and Benzoyl chloride.

Then it is heated upto 100°–140°C after bringing the temperature to room temperature. The process is called vulcanisaiton

Improvements or Objectives of Vulcanisation

- Linear chain molecules are converted to cross linked molecules

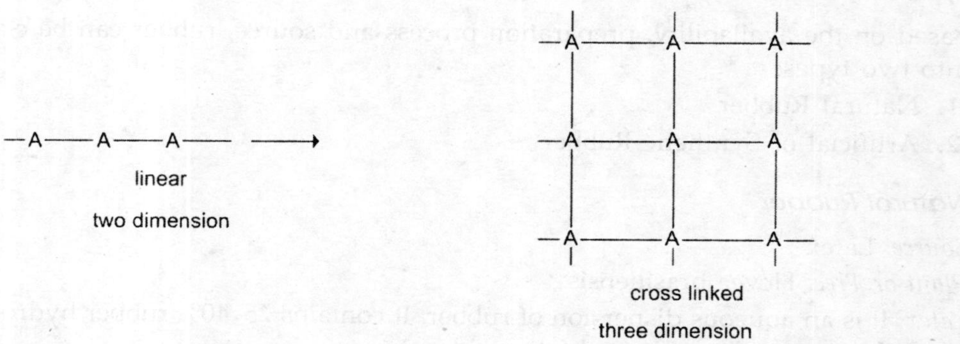

- Double bond is removed in this process
- Based on the sulphur addition, the following polymer characters are improved:
  1. Tensile Strength (Loading capacity)
  2. Elasticity
  3. Abrasion
  4. Chemical resistance
  5. Shelf life
  6. Stiffness
- Because of two dimension to three dimension structural changes, inter molecular movement is prevented completely

  Example for Vulcanised Rubber:

1. Tyre rubber —3 to 5% Sulphur content
2. Battery case Rubber —30% Sulphur content

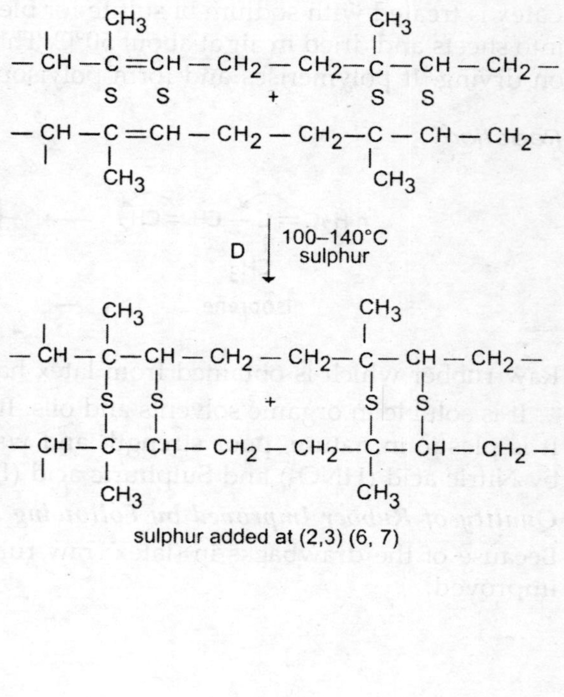

sulphur added at (2,3) (6, 7)

2.25 VULCANISED RUBBER

Uses of Rubber

- It is used to prepare tyres, toys and sports items, etc
- It is used to prepare heart valves and other stitching mesh, etc. in medicine
- Rubber gaskets are used in various equipments

  Example, Cookers and Engine, etc.

Differences between Raw Rubber and Vulcanised Rubber

No.	Raw Rubber	Vulcanised Rubber
1.	It is solube in organic solvents and oils	It is insoluble in organic solvents
2.	It absorbs water highly	Less water absorbing character
3.	Lower Tensile strength	Higher Tensile strength
4.	Lower oxidation resistivity	Higher oxidation resistivity
5.	It is affected by Nitric acid (HNO_3) and Sulphuric acid (H_2SO_4)	It is not affected easily
6.	Poor elasticity	Higher elasticity
7.	It must be used at low temperature 10–40°C	It can be used at 40–100°C
8.	Shelf life is less	Self life is more

2.26 SYNTHETIC RUBBER

Because of increased, requirement we have to look up to synthetic rubber. Vehicles are increased, forests are reduced, to avoid this problems synthetic rubbers are prepared. Synthetic polymers are artificial polymers. Artificial rubbers have better quality than natural rubber.

Example, for Synthetic Polymers

1. Buna - S Rubber or GR - S Rubber or SBR
2. Butyl Rubber.

Preparation of Buna - S or SBR or GR - S Rubber and SBR (Styrene Butadiene Rubber)

Type of Rubber: artificial or synthetic rubber.

Type of polymerization: Copolymerisation

Monomers: Butadiene, Styrene

Reason for name: Buna - S Bu - Butadiene, Na - Natrium (Sodium), S - Styrene

Reaction

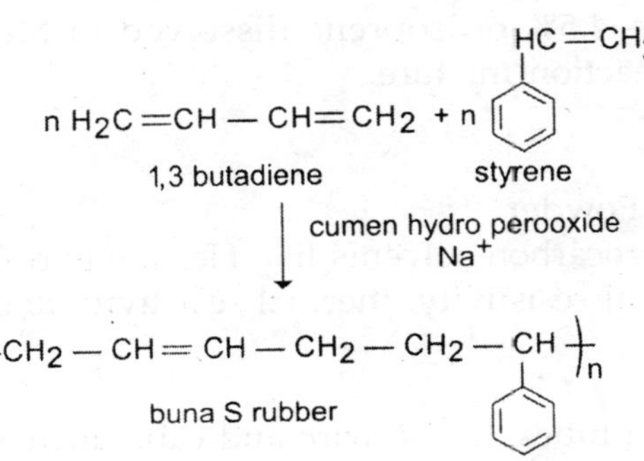

Process

75% Butadiene and 25% Styrene react together in the presence of sodium and cumene hydroperoxide as an aqueous solution.Cumene hydroperoxide acts as an emulsifying agent

Uses

For making Motor tyres, Gaskets, Gum and Hoses, etc.

Butyl Rubber

Type of Rubber: Artificial or Synthetic Rubber
Type of Polymerisation: copolymerization.
Monomers: Isobutylene, Isoprene
Catalyst: Anhydrous Aluminium Chloride (AlCl3)

Reaction

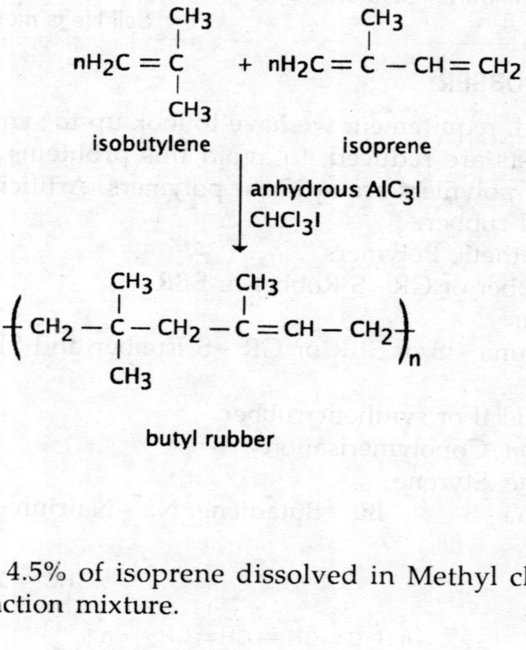

butyl rubber

Process

Isobutylene and 1.5 – 4.5% of isoprene dissolved in Methyl chloride.Anhydrous $AlCl_3$ added to the reaction mixture.

Properties

- It is a Amorphous Powder
- It is soluble in hydrocarbon solvents like Hexane and Cyclohexane
- It has good electrical resistivity, thermal resistivity and strength

Uses

Used for making cycle tubes, hoses, wire and cable insulation

2.27 COMPOSITES

Need for Composite

- All plastics and polymers are not suitable for some engineering uses
- To fulfill this requirement, new composition polymer was required to be developed
- To satisfy the specific requirements in industry and general purpose the developed material is called composites.
- Composites are immiscible polymers, e.g.
  Aircrafts (Supersonic) and Spacecrafts, etc.

Definition

Composite is a mixture of two or more macromolecules. Which are mutually insoluble and different composition and different phases.

Uses of Composites

- Composites are used to prepare bus top
- Plywood joined by composites
- Photo sheets are also a type of composite

Properties of Composites

i. Composites have higher
   a. Specific strength
   b. Specific stiffness
   c. Thermal resistivity
   d. Corrosion resistance
   e. Oxidation resistance
ii. Composites have lower
   a. Specific gravity
   b. Electrical conductivity
iii. It has good strength at high temperatures and it is easily fabricable

Constituents of Composite

Composites have two important (components) parts
  i. Matrix phase
  ii. Dispersed phase

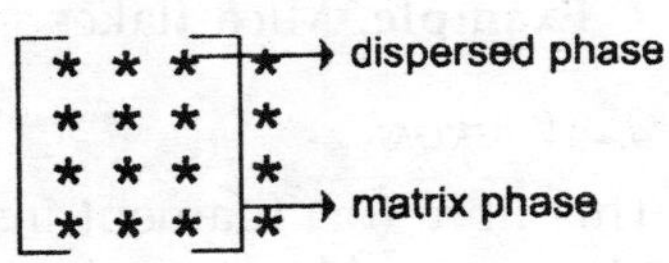

Matrix Phase

It is a continuous body constituent which is coated by a different metal or ceramics or polymer.

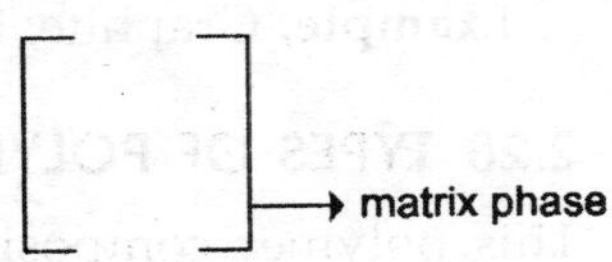

Types of matrix phases

1. Metal matrix phase
2. Ceramic matrix phase
3. Polymer matrix phase

Properties

Higher bonding strength between the matrix and dispersed phases

Function of Matrix phase

It binds the dispersed phase together. It transmits and distributes external load to the dispersed phase. It prevents brittle cracks and has plasticity and softness. It protects the dispersed phase from chemical reaction

Dispersed Phase

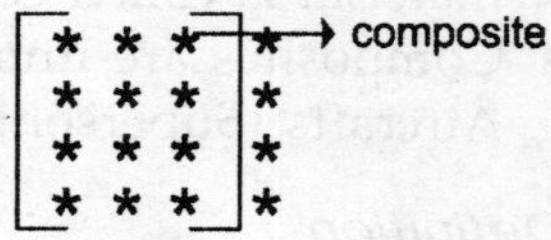

It is the internal structure of composite

Example,

1. Fibres
2. Particulates
3. Flakes
4. Whiskers

1. Fibres

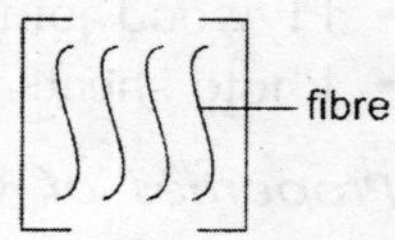

It is a long and thin filament of any polymer. It has higher length and high diameter ratio

If the matrix is unidirectional, the resulting composites are called anisotropic.

Example, Glass fibre, Carbon fibre and Aramid fibre

2. Particulates

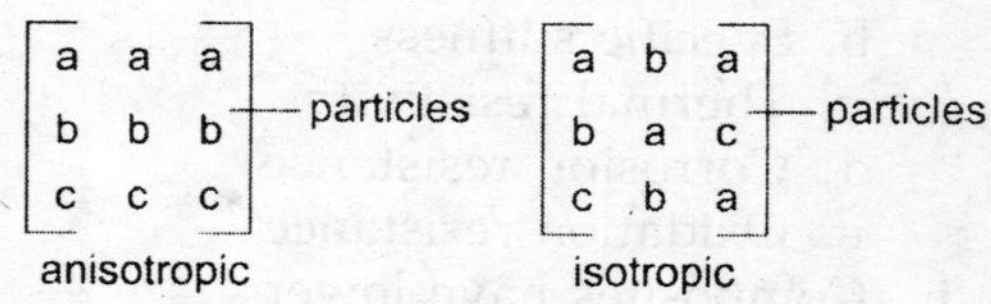

Matrix has random distributed phases and Small pieces of hard solid material. Pieces may be metallic or non- metallic. Random distribution of particles is called isotropic matrix phase.

Example, Iron metal powder, Glass powder

3. Flakes

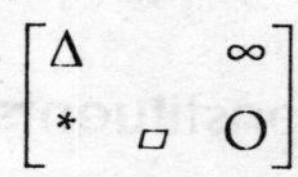

Thin solids that do not have exact structures are called flake. It has two dimensional solids (x,y axis only)

Example, Mica flakes

4. Whiskers

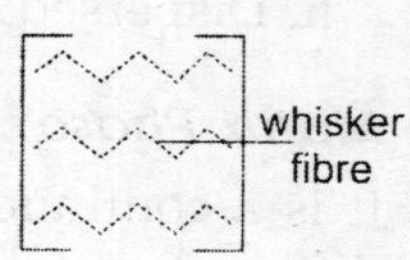

This fibre (or) filament inside the composite and Strong fibre (or) filament inside the composite. These all are prepared from growing crystals.

Example, Graphite, silicon Carbide (SiC) and Sapphire

2.28 TYPES OF POLYMER COMPOSITES

This polymer composites are called fibre reinforced polymer. Simply, it is called FRP,

Components of FRP (or) Fibre reinforced polymer

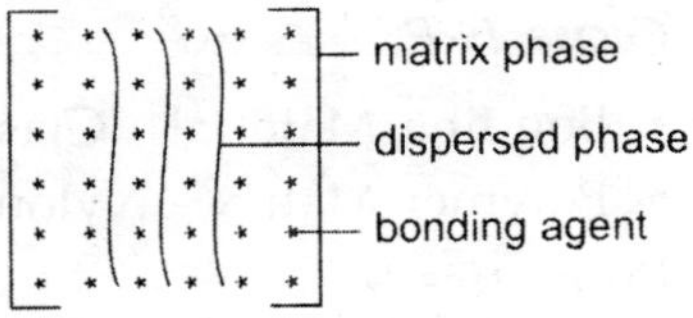

1. Filament (Dispersed)
2. Polymer Matrix
3. Bonding Materials

Preparation of FRP

Bonding the fibres with resin polymer matrix. Then it sticks under pressure and heat

Examples for Components:

1. Bonding or Sticking Material
   Example, Glass and Graphite
2. Filament or Fibre
   Example, Glass fibre and Metallic fibre
3. Polymer Matrix
   - Epoxy resin – Good Mechanical Strength
   - Polyester resin – Good Strength and Mechanical Property
   - Phenolic resin – Temperature resistivity
   - Silicon resin – Good thermal and Electrical resistant

Characteristics of FRP

It has a higher fatigue life. In FRP composites, the polymers mechanical properties are improved.

It has higher corrosion and heat resistance.

Types of FRP

Based on the bonding material it is classified

1. Carbon FRP
2. Glass FRP
3. Aramid FRP
4. Boron FRP
5. Alumina FRP

Carbon FRP or Carbon Fibre Reinforced Polymer

Properties

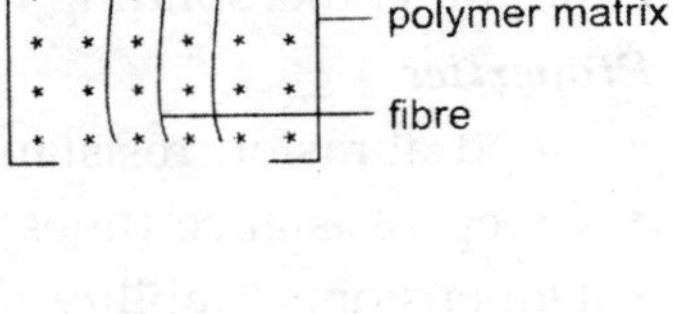

- This is high performance composites
- Low density
- Corrosion resistance
- Thermal resistance

Uses

Structural parts of aircraft and helicopter

Glass FRP

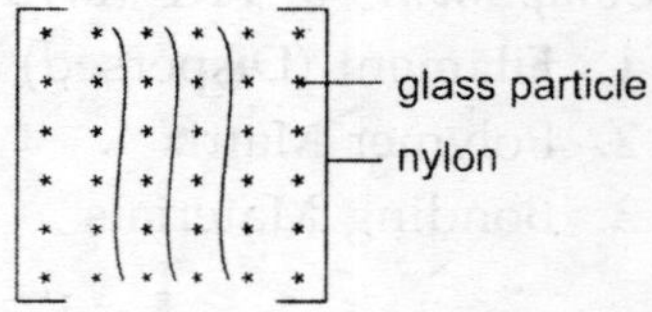

- Bonding Material - Glass
- Polymer Matrix - Nylon or Polyester

Properties

- Low density
- High tensile strength
- High corrosion resistance

Uses

It is used in Automobile parts, Storage Tank and Plastic pipe.

Aramid FRP

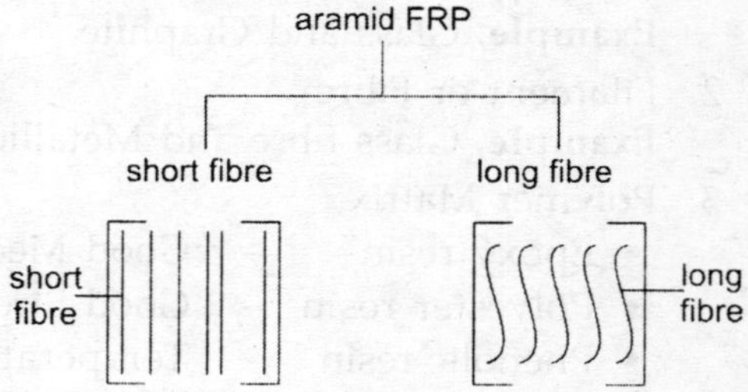

Based on the length of fiber, it is of two types:

Properties of Short Fibre Aramid:

- Effective reinforcement
- High surface area and strength
- High toughness

Uses

It is used to prepare Automobile breaks and Clutches preparation

Long Fibre

- Metal like fibre
- Ductile character
- Non-catastrophically (breaking character)
- It absorbs energy

Uses

In aircraft and helicopter parts are prepared by long fiber FRP.

Boron Fibre Reinforced Polymer (Boron FRP)

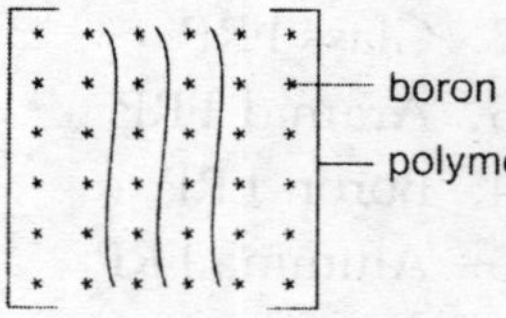

Boron is the bonding material.

Properties

- High stiffness and high strength

Uses

Aeroplane horizontal and vertical tails are prepared by using boron FRP.

Alumina FRP

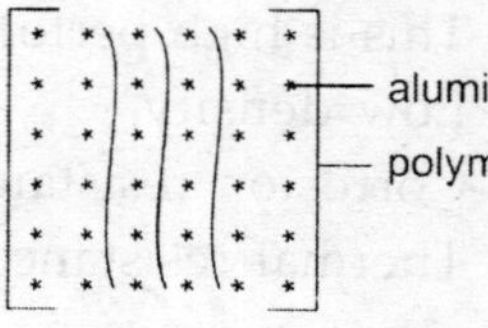

Alumina is the bonding material

Properties

- Good abrasion resistance
- Creep resistance (lines do not form)
- Dimensional stability

Uses

- Engine parts preparation
- Turbine engine parts are prepared by alumina FRP.

Importance of Polymer in Engineering Chemistry

- Electrical and Electronics spares and instruments are made up of polymer. To select the body of the instrument, polymer knowledge is important.
- Instrument circuit boards are made up of polymer material. To avoid the melting nature of circuit board, Technical people should know about polymer nature and to design and select the polymer for preparing instruments.
- To prepare the polymer vessels and polymer materials, machines are designed. Machines are designed by mechanical engineers. Vehicles spare parts are made up of polymer materials. For example, radiator circulated tube to made up of polymer material.
- Nowadays some polymers are used for strengthening the cement and bricks Polymer properties and purities are determined by instruments. Softwares are developed to correlate the properties. To develop the software, engineers should know about the polymer properties. Lot of embedded softwares are used in analytical instruments.

GLOSARY

Addition polymers: Those obtained by addition reaction between monomers.

Cationic polymerization: A type of polymerization involving a cationic site as an initiator.

Hetero polymer: If any polymer contains different atoms in chain polymer.

Homo polymer: If any polymer contains same atoms in chain polymer.

Plasticizers: A material that is added to resins to improve their flexibility.

Nylon: Polyamide polymer.

Veneer: A very thin sheet of polymer.

Thiokol: Polysulphide rubber.

Assessment Questions

Answer all questions (2 marks)

1. Define monomer.
2. Define polymer.
3. What is degree of polymerization?
4. What is polymerization?
5. Name the various polymerizations.
6. What are the various steps of free radical mechanism?
7. What is a dead polymer.
8. What is copolymerization? Give an example.

9. Distinguish between additional polymerization and condensation polymerization.
10. How is nylon 6:6 formed? Bring out its important properties and uses.
11. What are plastics?
12. What are the disadvantages of plastic?
13. List out the various ways by which polymers can be classified.
14. What are engineering plastics?
15. What are the important applications of high performance plastics?
16. Differentiate thermoplastic and thermosetting plastics.
17. Distinguish between commodity and engineering plastic.
18. Bring out the differences of general purpose plastics and engineering plastics
19. Mention preparation and uses of PVC
20. How is PVC prepared?
21. What is fluon? Mention its uses.
22. Mention some important application of polycarbonate.
23. Give two properties and uses of perlon-U.
24. How is perlon-U prepared?
25. What are elastomers?
26. Raw rubber cannot be used, why?
27. What is vulcanization of rubber?
28. What are the characteristics of FRP?
29. Draw the general structure of rubber
30. How is natural rubber obtained?
31. What are composites?
32. Write the characteristics of composites.
33. What are the constituents of composites?
34. How are composites classified?
35. What are FRPs?
36. Explain the properties of FRP?
37. Name some important FRPs.
37. Mention important applications of FRPs.
38. What is SBR?
39. Mention some important uses of SBR.
40. What is GR_1 rubber?
41. What are the characteristics of butyl rubber?
42. What is Teflon?
43. Which types of polycarbonates are prepared?
44. Explain perlon-U, PET, PC
45. Explain drawbacks of raw rubber.
46. Explain SBR.
47. Explain natural rubber.
48. Explain composites.
49. Mention properties and uses of perlon-U.

Answer all questions (16 Marks)

1. Distinguish between addition and condensation polymer with one example.
2. What is copolymerization? Give two examples.
   i. Define the term addition and condensation polymers. Give example.
   ii. Write the name of the monomers used in the preparation of nylon.
3. Mention its uses.
4. Write a note on the preparation, properties and uses of nylon 6:6.
5. Name any three engineering plastics and discuss about them.
6. Write the preparation, properties and uses of polycarbonate and Teflon.
7. Give an account of the preparation, properties and uses of PVC and Teflon.
8. Give the preparation and uses of the polycarbonate.
9. What are the characteristics of polycarbonate and polyurethane?
10. Distinguish between thermoplastics and thermosetting plastics.
11. Discuss the method of preparation and properties of Teflon.
12. State two uses of PVC.
13. Describe the preparation, properties and uses of polycarbonate and teflon.
14. Give the preparation, properties and application of polyamide.
15. What are drawbacks of raw rubber? How to improve the properties of rubber? Explain.
16. Compare the raw rubber with vulcanized rubber.
17. Explain the process of vulcanization.
18. Explain the mechanism of the free radical polymerization.
19. Explain the process of vulcanization of rubber.
20. Explain properties and uses of FRPs
21. Explain the various types of polymerization with examples.
22. Write preparation, properties and uses of (i) PET (ii) polycarbonate
23. Write about some engineering plastics with special characteristics.
24. Write preparation, properties and uses of polycarbonate and Teflon.
25. Write the mechanism of free radical polymerization.

3

Surface Chemistry

3.1 INTRODUCTION

This is an important technology for engineering students, using these techniques, water is purified. Metal surface nature is retained by this phenomenon, charcoal is used for a lot of engineering processes and pharmaceutical processes. It is based on surface chemistry. Weather report is also based on this surface phenomenon. It is very much of use in chemical industries. The simple processes of writing is also nothing but surface chemistry.

3.2 DEFINITION FOR SURFACE CHEMISTRY

The chemistry which explains the surface reactions and nature is called surface chemistry.

States of Matters

Matters are classified into three types: Solid, liquid and gas. So, surface chemistry deals with solid particle surface characters.

3.3 ADSORBATE

These are the surface seekers. It is stuck or adsorbed on the surface. These have dispersion characters.

Example, Gas or Liquid

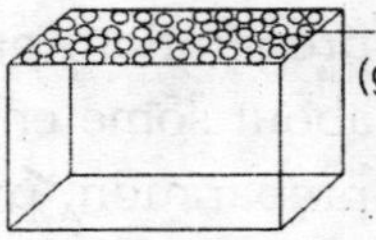

O_2, N_2, H_2 and H_2O

Fig. 3.1

3.4 SORPTION

In this, both adsorption and absorption occur one by one. Adsorption starts first and absorption continues

$$\text{Sorption} = \text{Adsorption} + \text{Absorption}$$

3.5 TYPES OF ADSORPTION

Based on the bond strength and force, adsorption can be classified into two types:

1. Physisorption
2. Chemisorptions

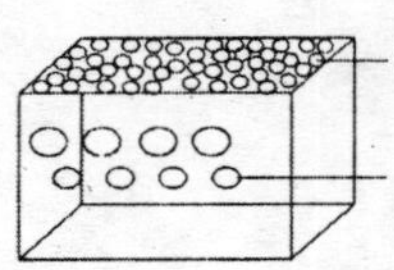

Fig. 3.2

Solution

Solution = Solvent + Solute

3.6 ABSORPTION

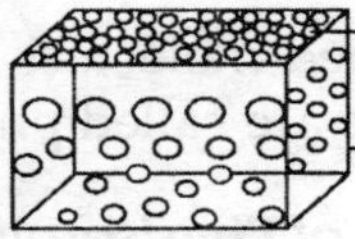

Fig. 3.3: Absorption

Gas or liquid molecules are spread over the solid body in absorption. In this process

Distribution is uniform throughout the body.

Chalk piece in water, Sponge in water, Ammonia in water and Water absorbed by anhydrous $CaCl_2$ are the example for an absorption.

Adsorption

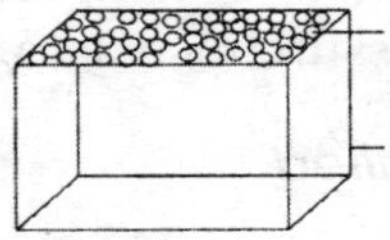

Fig. 3.4

The concentration of gas or liquid molecules gradually increases on solid surfaces in adsorption.

Charcoal adsorbs NH_3 (Ammonia), Starch adsorbs I_2 (Iodine) are a few examples for adsorption.

Adsorbent

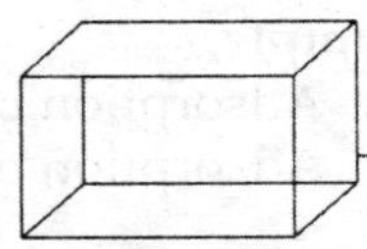

Fig. 3.5

It provides a surface for new molecule or gas, this solid surface is called an adsorbent.

Example, Charcoal, Clay, Alumina and Silica

Heat of Adsorption

During adsorption heat is evolved. The heat evolved for 1 unit(gram) mass adsorption is called heat of adsorption.

Occlusion

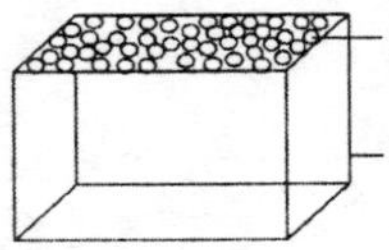

Fig. 3.6

Adsorption of gas molecules on solid surfaces is called occlusion.

Example, H_2 gas on charcoal.

Desorption

The reverse process of adsorption is called desorption. Removal of gas or liquid from solid surface is called desorption.

3.7 PHYSISORPTION

Physisorption is defined as the adsorption because of weak physical van der Waals bonds.

Example, H_2 or O_2 gas on charcoal.

3.8 CHEMISORPTIONS

The adsorption because of strong chemical covalent bond is called chemisorptions.

Example, O_2 gas adsorbs on metal surfaces, H_2 gas adsorbs on nickel surfaces

Physisorption and its Character

The adsorbed molecules (gas / liquid) on solid surfaces by weak physical van der Waals force are called physisorption.

Properties of Physisorption

- It is bound or adsorbed by weak forces
- It is a reversible process because of weak forces
- On solid surfaces mono or multi layers are formed
- Lower heat of adsorption
- Because of temperature increases, physisorption decreases
- Pressure increases, physisorption increases

Simplified

Physisorption α Pressure

$$\text{physisorption} \propto \frac{1}{\text{temperature}}$$

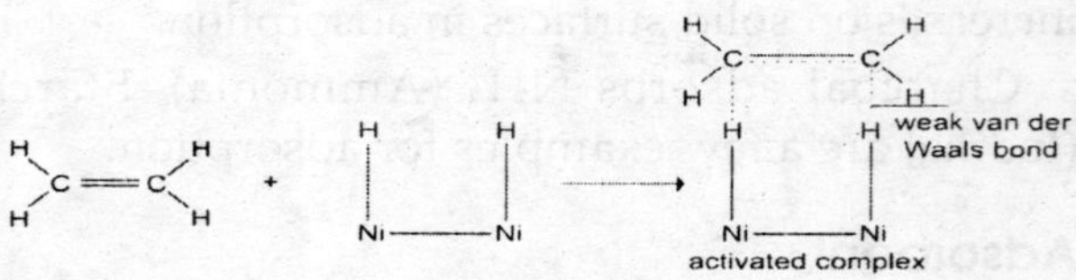

Fig. 3.7

Example,

1. Adsorption of H_2 on Nickel surface
2. Adsorption of H_2 on Palladium surface are the examples for physisorption.

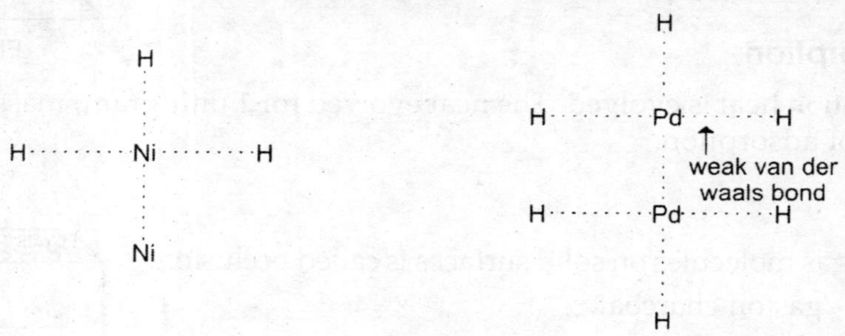

Fig. 3.8

Chemisorptions

The adsorbed molecules strongly bound by strong covalent bonds on solid surfaces is called chemisorptions.

Properties of Chemisorptions

- Strong covalent bonds are present between solid surfaces and the adsorbate (Gas or Liquid).
- The adsorbate forms a chemical compound on solid surfaces. This is called as surface complex.
- It emits high heat of adsorption and is a highly exothermic process.
- It is an irreversible process

- Mono layer formation occurs on solid surfaces
- Chemisorption is directly proportional to Temperature. Temperature increases, chemisorption increases.

Chemisorption α Temperature

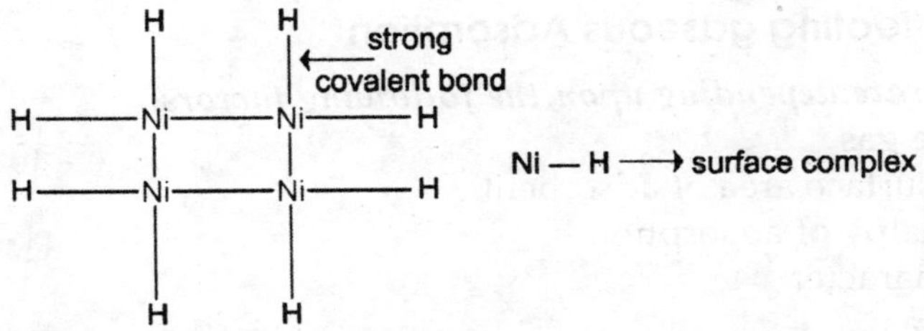

3.9 DIFFERENCES BETWEEN PHYSISORPTION AND CHEMISORPTIONS

No.	Physisorption	Chemisorption
1.	Physisorption is because of weak physical van der Waals bonds	Chemisorption is because of strong covalent bond
2.	It is a reversible process	It is an irreversible process
3.	At lower temperatures adsorption occurs	At a higher temperature adsorption occurs
4.	Mono or multilayer formation on solid surfaces	Mono layer forms on solid surfaces
5.	Temperature increases, adsorption decreases Ads $\alpha \frac{1}{T}$	Temperature increases, adsorption increases Ads α T
6.	Heat of adsorption is low (20–40 kcal/mol)	Heat of adsorption is high (40–400 kcal/mol)
7.	Pressure increases, adsorption increases, Ads α P	Ads $\alpha \frac{1}{P}$
8.	Concentration of adsorbate increases. Physisorption increases Ads α C	Decreases with adsorbate concentration Ads $\alpha \frac{1}{C}$
9.	Adsorbate Volume > Adsorbent Volume	Adsorbate Volume = Adsorbent volume
10.	Small activation energy	Higher activation energy
11.	Equilibrium reaches fast	Slowly reaches an equilibrium
12.	No chemical reaction	Chemical reaction occur
13.	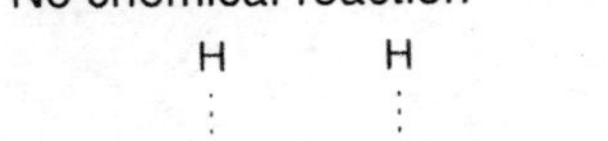	

3.10 BASED ON THE ADSORBATE, ADSORPTION IS CLASSIFIED INTO TWO TYPES

1. Adsorption of Gases on Solid
2. Adsorption of Solute from Solution

We can define solution by the following simple equation

Solution = Solvent + Solute

Type 1

Adsorption of Gases on Solid

Adsorption of gases on solid surface is by physisorption or chemisorptions.

3.11 Factors Affecting gaseous Adsorption

This adsorption rate depending upon the following factors.

1. Nature of the gas
2. Nature and surface area of adsorbent
3. Heat or Enthalpy of adsorption
4. Reversible character
5. Pressure of gas
6. Temperature of gas
7. Thickness of the adsorbed gas layer

Nature of Gas

Liquefaction: Gaseous molecules are converted to liquid in the process of liquefaction. HCl, NH_3, SO_2 etc. are examples of liquifiable gases. It has a higher critical temperature
Permanent gases: H_2, N_2, O_2 are examples of permanent gases. It has a less critical temperature. Liquefied gases are adsorbed easily than permanent gases.
Reasons for the above statement:

- Liquefied gases are bonded by weak Vander walls bonds (Physisorption) and permanent gases are bonded by strong covalent bond that's why liquefied gases are adsorbed easily than permanent gases.
- Critical temperature

  At which particular temperature, the gaseous state is condensed to the liquid state is called critical temperature. Critical temperature increases, adsorption also increases.

Adsorption of gases on 1 gm Activated Charcoal is tabulated.

Gases	CH_4	CO_2	H_2	N_2
Critical temperature °K	190	304	33	126
Amount of gas adsorbed (cm^3)	16	48	4.5	8

Nature and Surface area of adsorbent

Absorbent (solid) surface area increases, adsorption also increases. For example, cubic solid structure has 6 surfaces. If it is divided into 4 pieces. It has, 6 x 4 = 24 surfaces. If the absorbent has large pores, then only the rate of adsorption will be increases.

Example:

- Adsorption of H_2 on charcoal (more pores)
  Adsorption of H_2 ion nickel fine powder
- Adsorption of moisture and gas on silica gel

Heat of Adsorption

Adsorption is an exothermic reaction or heat emission reaction. Heat liberated when one gram of gas is adsorbed on solid surfaces is called heat of adsorption. During physisorption, process very small amount of heat is evolved, but in the chemisorption process, more amount of heat is evolved.

Reversible Character

Phsisorption is a Reversible process because in this process, surface compound is not formed and weak van der Waals bonds act between adsorbent and adsorbate, but chemisorption process is an irreversible process due to the formation of surface compounds on surfaces and covalent bond acts between adsorbent and adsorbate.

Pressure of Gas

There is an equilibrium between gas and solid surfaces. When Pressure increases, adsorption also increases. Pressure decreases, desorption occurs. So, adsorption is directly proportional to pressure. But for the comparison of physisorption and chemisorption, physisorption is directly proportional to pressure and chemisorptions inversely proportional to pressure.

Temperature

Temperature decreases, adsorption increases. It is inversely proportional to temperature. Physisorption is inversely proportional to temperature.

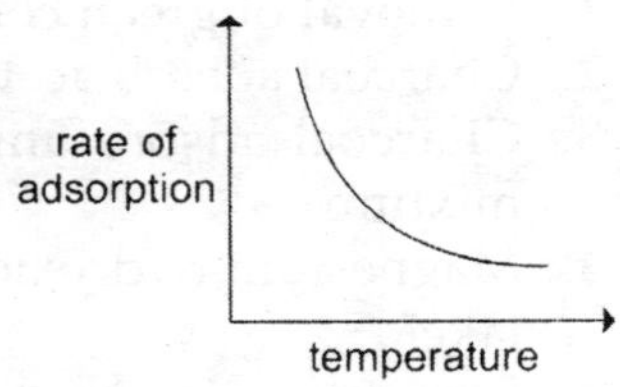

Fig. 3.9

$$\text{physisorption} \; \alpha \; \frac{1}{T}$$

Chemisorption is directly proportion to temperature

$$\text{chemisorption} \; \alpha \; T$$

Example, Physisorption: N_2 on iron at 83°K

Chemisorption: N_2 on iron at 775°K, N_2 forms Nitrides on iron surface.

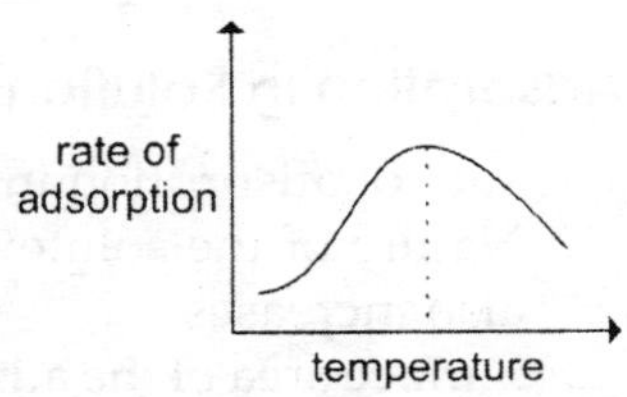

Fig. 3.10

Thickness of Adsorbed Layer of Gas

Pressure is directly proportional to adsorption. Pressure increases, multilayer is formed on solid surface.

Lack of Specificity

In physisorption, van der waals forces are universal. They do not show any specificity. A solid surface does not show very strong attraction for a few particular gases. Physisorption has low (lack of) specificity and Chemisorption has high specificity.

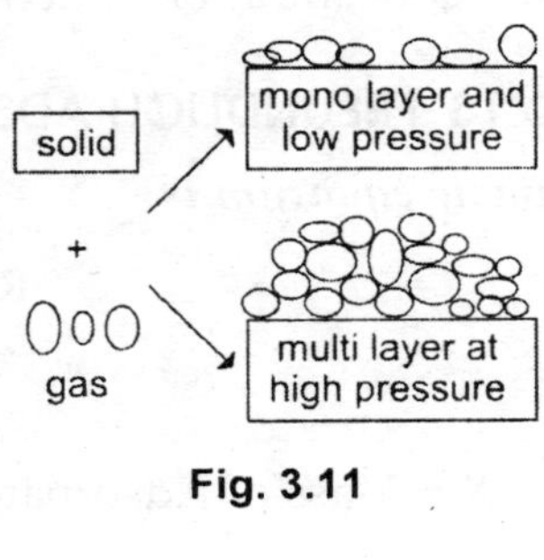

Fig. 3.11

Type 2

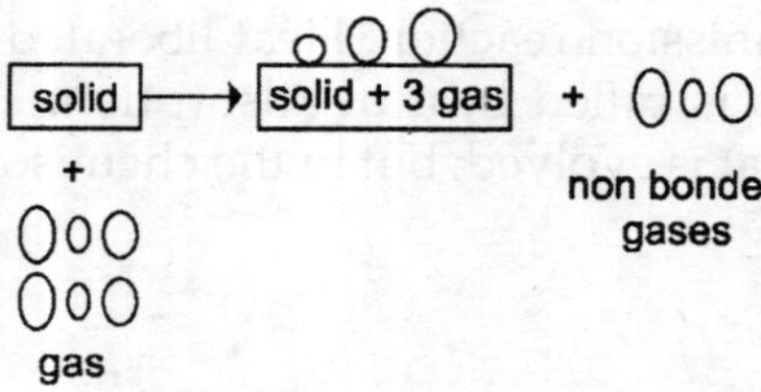

Fig. 3.12

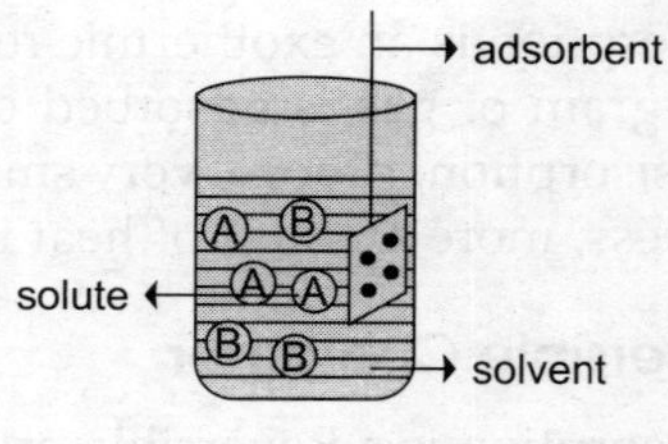

Fig. 3.13

3.12 ADSORPTION OF SOLUTES FROM SOLUTION

Absorbent - solid, Solute - dissolved material, Solvent - dissolving material
Solution is a mixture of solvent + solute

Adsorption of Solute

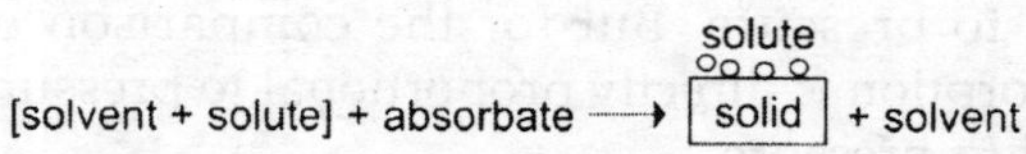

Example

1. Removal of green color from sugar solutions by charcoal
2. Charcoal adsorb acetic acid and oxalic acid in H_2O
3. Charcoal adsorb ammonia from ammonium hydroxide and phenolphthalein mixture
4. Magnesium hydroxide [Mg $(OH)_2$] adsorbs the dye magnes. This is called "Blue lake"
5. Sometimes in an electrolytic solution, alumina adsorbs electrolyte. Charcoal adsorbs non-electrolyte

Adsorption in Solution

The rate of adsorption in solution depends upon the following factors

1. Nature of the solute: Molecular weight of the solute is higher, solute adsorption also increases
2. Surface area of the adsorbent: Rate of adsorption increases as surface area increases.
3. Effect of temperature and Concentration: The relation between temperature and concentration is defined by Freundlich adsorption isotherm.

3.13 FREUNDLICH ADSORPTION ISOTHERM

Basic equation is

$$\text{Rate of adsorption} = \frac{\text{mass of adsorbate}}{\text{mass of adsorbent}} \quad \ldots 1$$

$$= \frac{x}{m} \quad \ldots 2$$

X = Mass of Adsorbate, m = Mass of Adsorbent

Rate of adsorption of gas α Pressure applied

$$\frac{x}{m} \propto P^{1/n} \qquad \ldots 3$$

If it is gas, pressure "*P*" term is used in the equation. For solute adsorption the equation should be in terms of concentration.

$$\frac{x}{m} \propto C^{1/n}$$

Derivation

$$\frac{x}{m} \propto C^{1/n} \qquad \ldots 4$$

$$\frac{x}{m} = KC^{1/n} \qquad \ldots 5$$

K – Adsorption Constant, N – Constant

Take "log" of both sides

$$\log \frac{x}{m} = \log kC^{1/n} \qquad \ldots 6$$

$$\log \frac{x}{m} = \log k + \log C^{1/n} \qquad \ldots 7$$

$$\log \frac{x}{m} = \log k + 1/n \log C \qquad \ldots 8$$

$$y = mx + C \qquad \ldots 9$$

$$y = \log \frac{x}{m};\ m = 1/n;\ x = \log C;\ \log k = C$$

Equation 9 is a straight line equation

Graph

Freundlich Adsorption.equation

$$\log \frac{X}{M} = \log k + \frac{1}{n} \log c$$

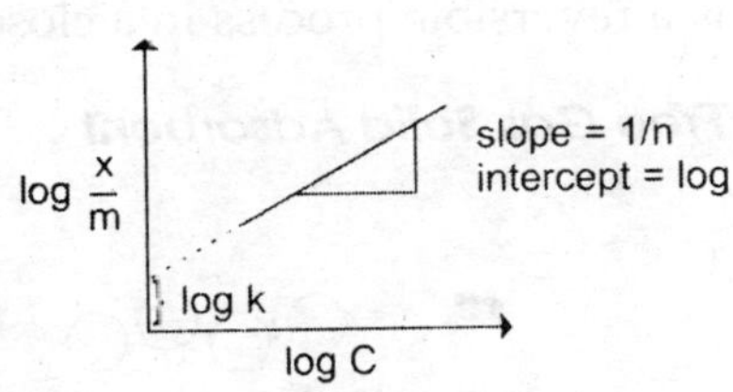

Fig. 3.14

If adsorbent adsorbs solute from the solution. This is called Positive adsorption. If the adsorbent adsorbs solvent from the solution. This is called negative adsorption.

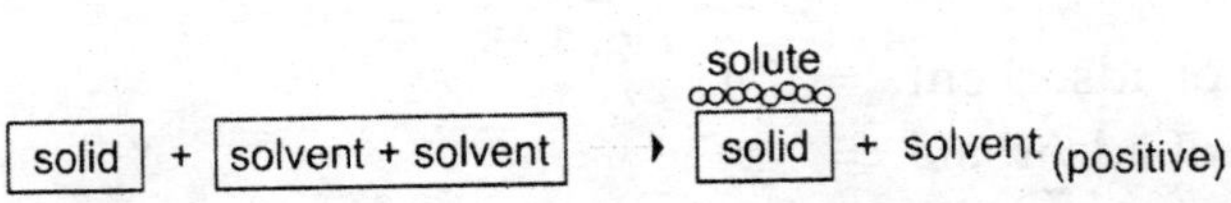

Example: From the concentration solution of NaCl, charcoal adsorbs NaCl from water.

Example: From dilute Salt solution Water adsorbed by charcoal

3.14 ACTIVATION OF ADSORBENT

To increase the rate of adsorption, adsorbent surface should be activated.

Adsorbent is activated by the following methods

i. By mechanical scratching the surface:

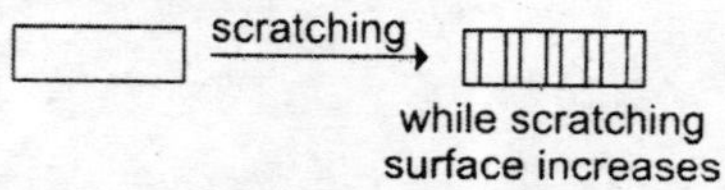

ii. Surface of the adsorbent is increased by making fine powder

iii. Pores are increased by heating process

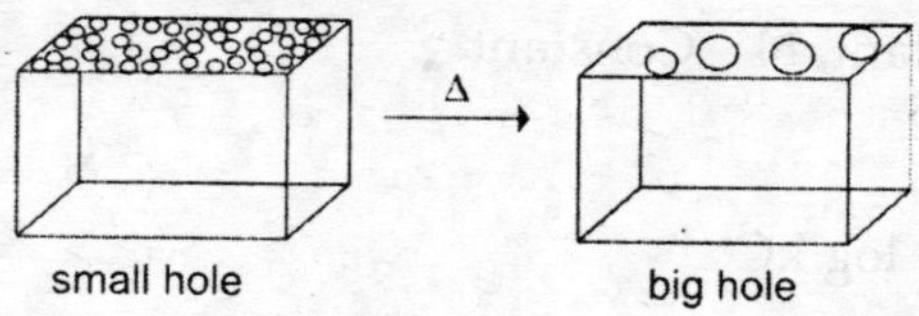

Because of heat energy molecules expand

3.15 ADSORPTION ISOTHERMS

The graphical representation of pressure and rate of adsorption per unit volume or unit mass of adsorbent at a constant temperature is called adsorption isotherm.

Adsorption Isotherm

Pressure Vs Adsorption per unit volume at a constant temperature

The adsorption of gaseous molecules on a solid surface is a reversible process in a closed container.

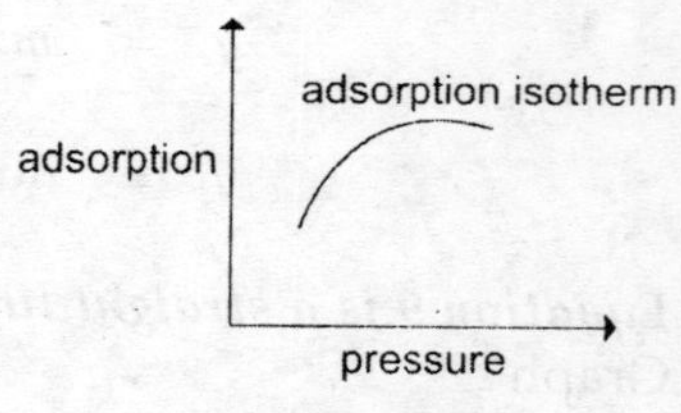

Fig. 3.15

Free Gas Solid Adsorbent

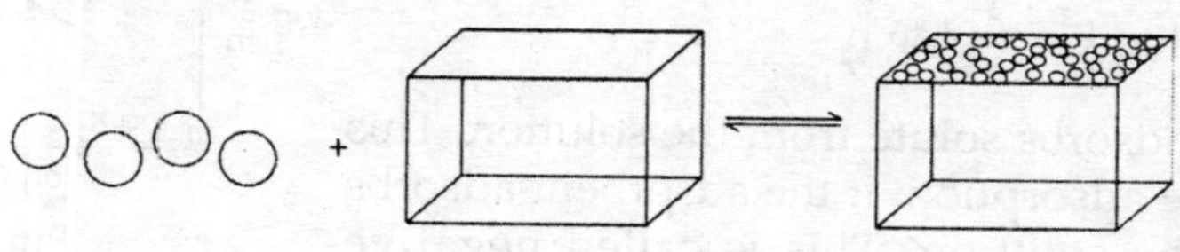

Fig. 3.16

Mass of adsorbent $= m$

Mass of adsorbate $= x$

Rate of adsorption $= x/\text{m}$

Rate of adsorption depends on Pressure

Saturation Pressure: At a particular pressure, adsorption is maximum and adsorption and desorption are at an equilibrium. This particular pressure is called saturation Pressure. After saturation pressure, adsorption is constant

$$\text{Adsorption maximum} \xrightleftharpoons{Ps} \text{Desorption}$$

Mathematical Expression: $\frac{X}{M} \propto P^{1/n}$

Graphical Representation:

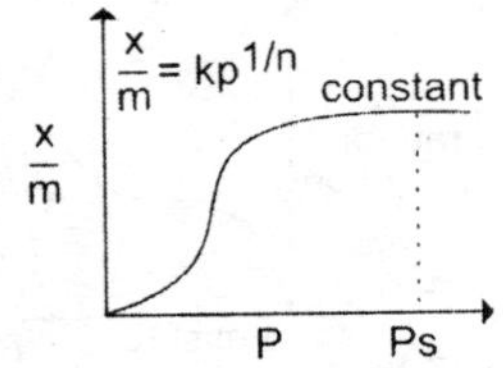

Fig. 3.17: Freundlich adsorption isotherm

Types of Adsorption Isotherms

Type I

Property: This isotherm explains Physical adsorption of Porous adsorbent and mono layer formation on the surface. For this adsorption, low pressure is enough.

N_2 gas adsorption on charcoal at 195°C, H_2 gas adsorption on Charcoal at 180°C are examples for the type I isotherm.

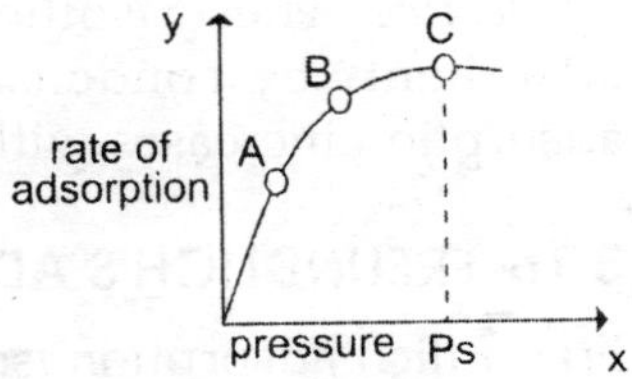

Fig. 3.18

From the graph,

Adsorption < Desorption at point A, Adsorption > Desoption at point B.

Adsorption = Desorption at a particular point C and adsorption is constant above the point "C"

Type II

Properties

This isotherm explains Physical adsorption of non Porous adsorbent and multi layer formation on the surface. For this type, adsorption increases with increasing pressure.

Adsorption of N_2 gas on Silica gel surface at 195°C, N_2 gas on Fe or Pt surface at 195°C are the examples for this type II adsorption.

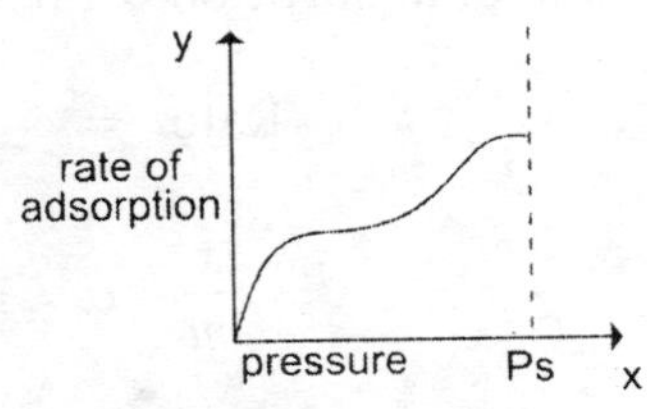

Fig. 3.19

Type III

Properties: This isotherm explains physical adsorption of porous adsorbent and a multi layer formation on the surface. For this type adsorption increases with increasing pressure.

Example, Adsorption of Br_2 on silica at 80°C

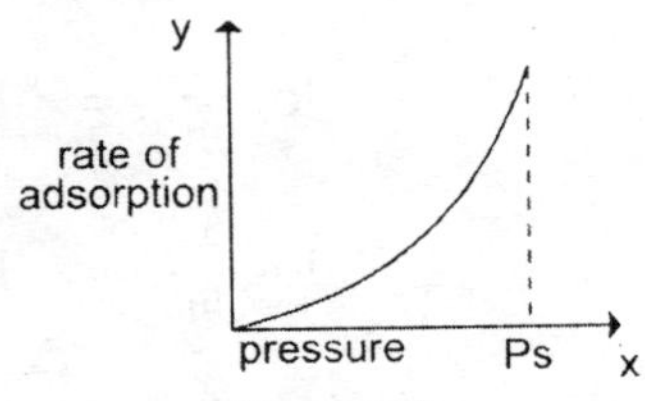

Fig. 3.20

Type IV

Example, Benzene on silica gel surface at 50°C

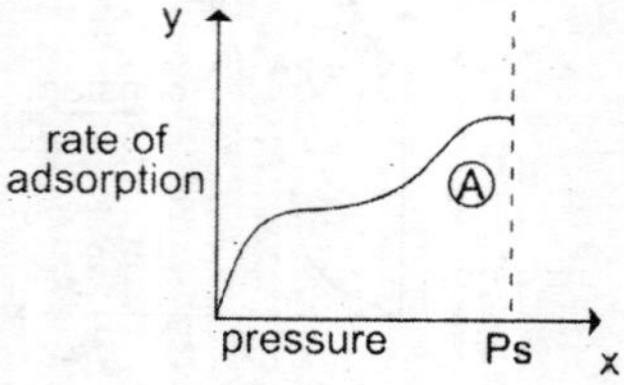

Fig. 3.21

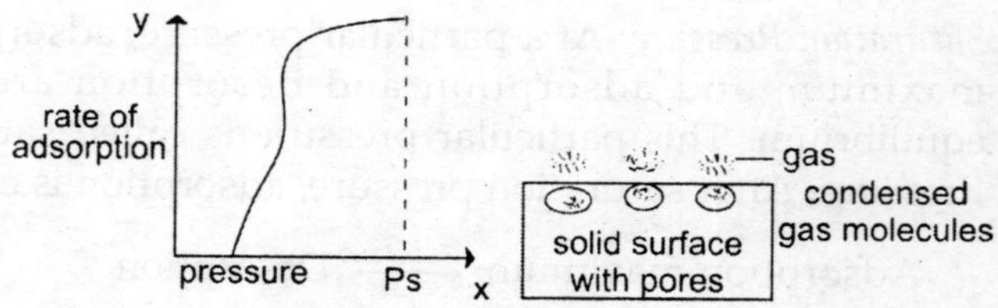

Fig. 3.22

Type V

Example, Water vapour on Charcoal surface at 100°C

Properties: These isotherms explain multi layer physical adsorption of porous adsorbents by condensation of gaseous molecule on the surface. For this type adsorption increases with increasing pressure.

3.16 FREUNDLICH'S ADSORPTION ISOTHERM

Freundlich adsorption isotherm for gaseous molecules adsorbed on a solid surface.

Principle: Rate of adsorption and Pressure at constant temperature is explained by Freundlich adsorption isotherm

Derivation

Rate of adsorption α Pressure$^{1/n}$

$$\text{Rate} = \frac{\text{Mass of adsorbate}}{\text{Mass of adsorbent}}$$

$$\frac{x}{m} \;\alpha\; P^{1/n} \qquad \ldots 1$$

$$\frac{x}{m} = KP^{1/n} \qquad \ldots 2$$

Take both side logarithm

$$\log\frac{x}{m} = \log(KP^{1/n}) \qquad \ldots 3$$

$$\log\frac{x}{m} = \log K + \log P^{1/n} \qquad \ldots 4$$

$$\log\frac{x}{m} = \log P^{1/n} + \log K \qquad \ldots 5$$

$$\log\frac{x}{m} = 1/n \log P + \log K \qquad \ldots 6$$

straight line equations is Y = mx + C ... 7

$$\log\frac{x}{m} = 1/n \log P + \log K \quad \text{... 8}$$

x-Mass of the adsorbate, m-mass of the adsorbent, p-pressure, k-adsorption constant, n-layeron surface.

Isotherm Graphical Representation

Properties of a graph: two types of graphs are formed in this process based on the pressure.

At lower pressure, a straight line is formed and at higher pressure, a slightly deviated curve (slight) is formed.

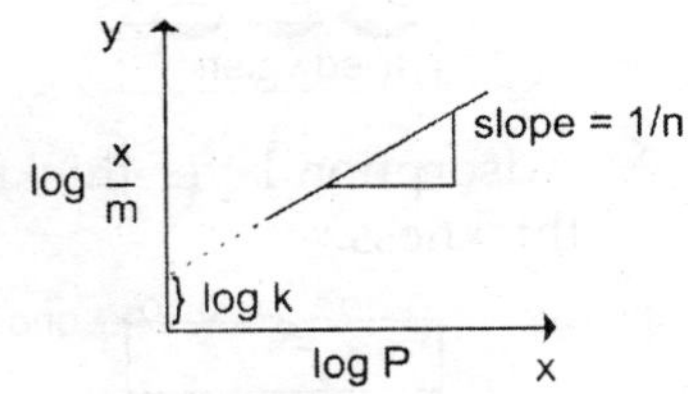

Fig. 3.23

Observations

1. *At low pressure:*

$$\frac{x}{m} = KP^{1/n}$$

where $n = 1$ low pressure

$$\frac{x}{m} = KP^{1/n}$$

$$= KP^1 = KP$$

2. *At high pressure:*

$$\frac{x}{m} = KP^{1/n}$$

where $n = 1$ high pressure

$$\frac{x}{m} = KP^0$$

$$= K \times 1 = K$$

adsorption constant at high pressure

3. *At medium pressure:*

$$\frac{x}{m} = KP^{1/n}$$

$$n = 0 \text{ to } 1$$

This is Freundlich isotherm

Limitations of Freundlich's Adsorption Isotherms

1. It is purely an empirical equation
2. This adsorption isotherm is invalid at high pressure.
3. Constants k and n are temperature dependent, they are not explained properly.
4. Equation is not explained and mentioned for higher concentration of adsorbate

3.17 LANGMUIR'S ADSORPTION ISOTHERM

Postulates:

To rectify the drawbacks of the Freundlich adsorption isotherm, Langmuir explained some postulates:

1. The valency of surface is not fulfilled.

2. Adsorption layer thickness on the surface is considered as of only one molecule thickness.

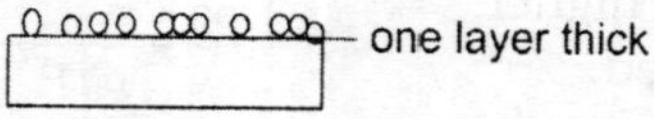

3. Surface of the solid is homogeneous in nature.

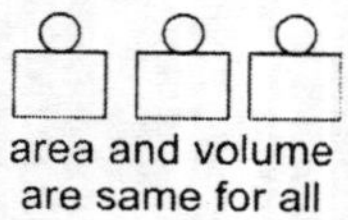

4. There is no interaction between adjacent molecules.

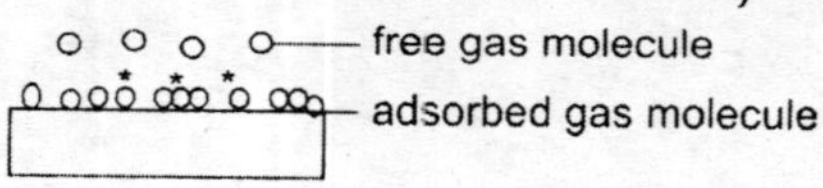

5. Adsorbed gas molecule does not move on surface.

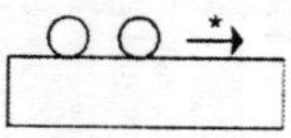

Derivation of Langmuir's Adsorption Isotherm

$$A(g) + M_{\text{solid}} \underset{\text{desorption Kd}}{\overset{\text{adsorption Ka}}{\rightleftharpoons}} A/M$$

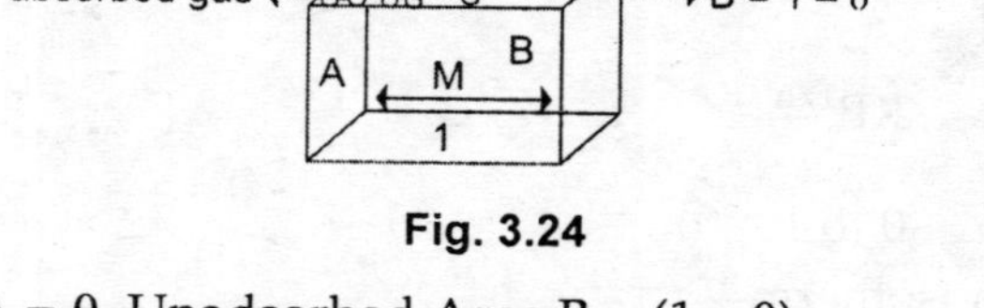

Fig. 3.24

Adsorption Area A = θ, Unadsorbed Area B = $(1 - \theta)$

Dynamic equilibrium between

$$\text{Adsorption} \longleftrightarrow \text{Desorption}$$

Rate of Adsorption = R_a
Rate of Desorption = R_d

Rate of Adsorption α Uncovered area × Pressure

$$R_a \propto (1-\theta)\,P \quad \ldots 1$$

$$R_a = K_a (1-\theta)\,P \quad \ldots 2$$

(K_a – adsorption constant)

Rate of desorption R_d α covered area

$$R_d \propto \theta \quad \ldots 3$$

$$R_d = K_d\,\theta \quad \ldots 4$$

(K_d – desorption constant)

At equilibrium: Rate of adsorption = Rate of desorption

$$R_a = R_d \quad \ldots 5$$

$$K_a (1-\theta)\,P = K_d\theta \quad \ldots 6$$

$$K_aP - K_a\,\theta\,P = K_d\theta \quad \ldots 7$$

Rearrange the equation

$$K_aP = K_d\theta + K_a\theta\,P \quad \ldots 8$$

RHS θ commonly taken out

$$K_aP = \theta(K_d + K_aP) \quad \ldots 9$$

RHS multiple come to the LHS, dividing

$$\theta = \frac{K_aP}{K_d + K_aP} \quad \ldots 10$$

To the above equation, the RHS numerator and denominator are divided by K_d

$$\theta = \frac{K_aP/K_d}{(K_d + K_aP)/K_d} \quad \ldots 11$$

$$\theta = \frac{(K_a/K_d)P}{(K_d/K_d) + (K_a/K_d)P} \quad \ldots 12$$

$$\frac{K_a}{K_d} = \frac{\text{adsorption constant}}{\text{desorption constant}}$$

$$\frac{K_a}{K_d} = K_{(\text{rate of adsorption constant})}$$

$$\theta = \frac{KP}{1 + KP} \quad \ldots 13$$

Volume of the adsorbed gas α adsorption covered area

$$X \alpha \theta \quad \ldots 14$$

$$X = A\theta \quad \ldots 15$$

Substitute θ value in the 15th equation

i.e., Constant A

$$X = \frac{AKP}{(1+KP)} \quad \ldots 16$$

$$(1 + KP) = \frac{AKP}{X} \quad \ldots 17$$

(cross multiplication of 16 th)

Both sides are divided by 'AK' in Equation 17

$$\frac{(1+KP)}{AK} = \frac{AKP}{AKX} \quad \ldots 18$$

$$\frac{1}{AK} + \frac{KP}{AK} = \frac{P}{X} \quad \ldots 19$$

Rearrange the 19th equation

$$\frac{P}{X} = \frac{K}{AK} \times P + \frac{1}{AK} \quad \ldots 20$$

P – Pressure, *K* – rate of adsorption, *X* – volume of adsorbate, *A* – constant.

Compare with straight line equation

$$Y = mx + C$$

$$Y = P/x; \quad m = K/Ak, \quad x = P, C = 1/AK$$

Equation 20 is Langmuir adsorption isotherm

Graph:

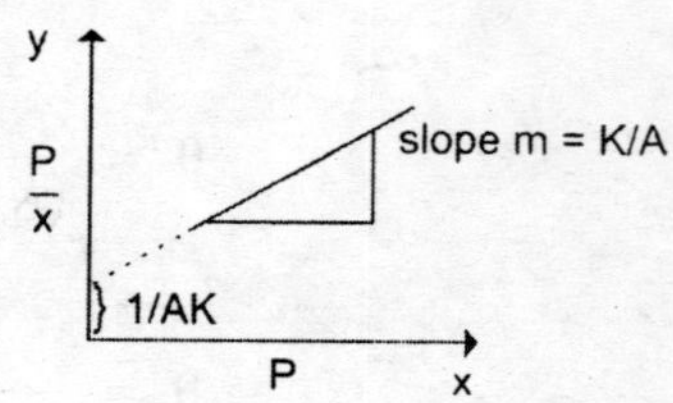

Fig. 3.24

Case Studies

$$\frac{P}{X} = \frac{K}{AK}(P) + \frac{1}{AK}$$

Case 1 (At low pressure)

In this equation, three physical parameters are present. Because of low pressure, pressure term must be negligible.

Term 2 was removed in the equation,

$$\frac{P}{X} = \frac{1}{AK}$$

Cross multiplication,

$$X = AK\,P;$$

$$X\,a\,P \qquad \ldots 21$$

Case 2 At High pressure

Term 3 was removed in the equation, then the equation becomes

$$\frac{P}{X} = \frac{K}{AK}(P) + \frac{1}{AK}; \qquad \frac{KP}{AK} > \frac{1}{AK}$$

$$\frac{P}{X} = \frac{KP}{AK}$$

Cross multiplication,

$$X = AP^0 \qquad \ldots 22$$

At high pressure, adsorption is completed

Case 3 Normal Pressure

At normal pressure the equation 23 becomes

$$X = AP^n \qquad \ldots 23$$

Draw back: Langmuir adsorption isotherm works at low pressure. pressure power becomes zero at high pressure, then volume of gas becomes constant

3.18 APPLICATIONS OF ADSORPTION TECHNOLOGY OR SURFACE PHENOMENA

1. *Charcoal:* In chemical plants and near to the industrial areas the air contain harmful and toxic gases. To avoid the inhaling of gas, masks are prepared with activated charcoal and air is purified by adsorption. For example, HCl vapours and ammonia gas adsorb on charcoal surface.
2. Charcoal is used to remove the color in many industries, for an example charcoal is used in:
   Pharmaceutical industries: The unwanted color in major products are removed by stirring with charcoal. In sugar industries chlorophyll (green) and other unwanted gases are removed by charcoal treatment. In an acetic acid(Vinegar) industries unwanted colour is removed by activated charcoal. For Purification of water,it is passed through charcoal.
   During this process Organic particles and floating solid particles are adsorbed by charcoal

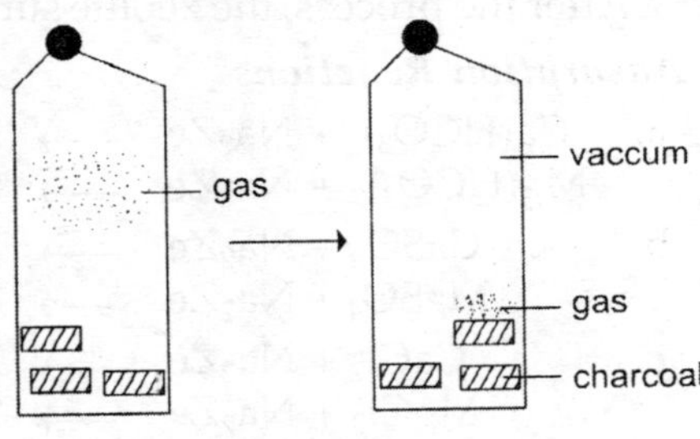

Fig. 3.25

3. To create vacuum in Dewar Flask charcoal and some other adsorbents are used.
4. Silica gel and Alumina gel (SiO_2 and Al_2O_3) are used to adsorb moisture in gas or air. In chemical instruments, N_2, H_2 and O_2 gases are purified in a weighing balance

5. *General Chemicals:* Arsenic poison gets adsorbed by Ferric hydroxide [$Fe(OH)_3$)]. Fullers earth (silicate polymer) is used to remove the unwanted material in petroleum products. This surface phenomenon is used in organic chemical reactions.
   Example, Reduction Adsorption Catalysts:
   a. Pd / C - Palladium and Carbon
   b. Ni, Pt - Nickel and Platinum
   c. Heterogeneous Catalysts
6. Adsorption technology is used in power plants to adsorb the excess neutrons in the nuclear reactor, mordant is used in the reactor. The dyeing process is also based on the adsorption phenomenon.

Total Applications of Surface Phenomena:

- Water purified is by the ion exchange method
- Sodium zeolite process
- Reduction reaction of alkene
- Activated carbon to purify air and water

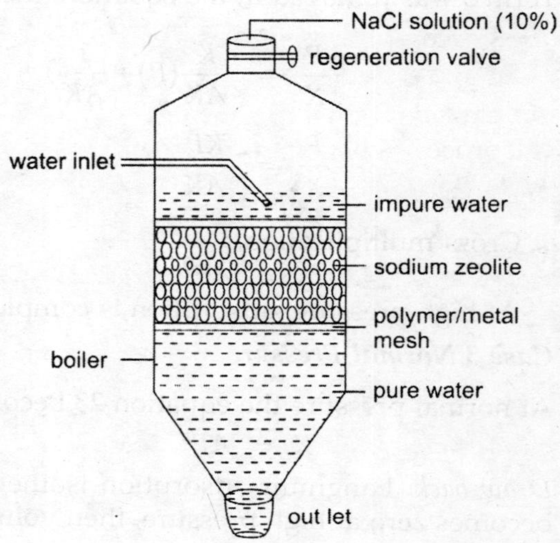

Fig. 3.26

3.19 ZEOLITE PROCESS

Process: Sodium zeolite process (Na_2Ze)
Technology: Adsorption Technology
Use: Purification of water
Phenomenon: Surface Phenomenon

Process

Sodium zeolite resin is filled inside the boiler. Raw water is passed through the boiler. Impure water contain Ca^{2+}, Mg^{2+} salts. These salts are removed by this process. Cations are adsorbed on the zeolite surface. Pure water comes out and is collected. After the process, the zeolite surface is regenerated by brine (sodium chloride) wash.

Adsorption Reactions

a. $$Ca(HCO_3)_2 + Na_2Ze \longrightarrow 2NaHCO_3 + CaZe$$
$$Mg(HCO_3)_2 + Na_2Ze \longrightarrow 2NaHCO_3 + MgZe$$

b. $$CaSO_4 + Na_2Ze \longrightarrow Na_2SO_4 + CaZe$$
$$MgSO_4 + Na_2Ze \longrightarrow Na_2SO_4 + MgZe$$

c. $$CaCl_2 + Na_2Ze \longrightarrow 2NaCl + CaZe$$
$$MgCl_2 + Na_2Ze \longrightarrow 2NaCl + MgZe$$

Regeneration Reaction

The surface of resin is washed by 10% NaCl (solution)

$$CaZe + 2NaCl \longrightarrow CaCl_2 + Na_2Ze$$
$$MgZe + 2NaCl \longrightarrow MgCl_2 + Na_2Ze$$

3.20 ION-EXCHANGE RESIN ADSORPTION

Ion exchanger

Synthetic Resin or Polymer. Exchangers have one adsorbed ion and insoluble organic compounds in it. It is called ion-exchange resin.

Ion Exchange Adsorption

The process where in released ions from the surface are adsorbed by other ions is called ion - exchange adsorption

Classification of Ion-exchangers

Based on the adsorbed ion nature

i. Cation exchangers
ii. Anion exchanger

Cation Exchanger

Technology: Adsorption Technology
Phenomena: Surface Phenomena
Name of the Exchanger: Styrene—Divinyl benzene copolymer
Structure:

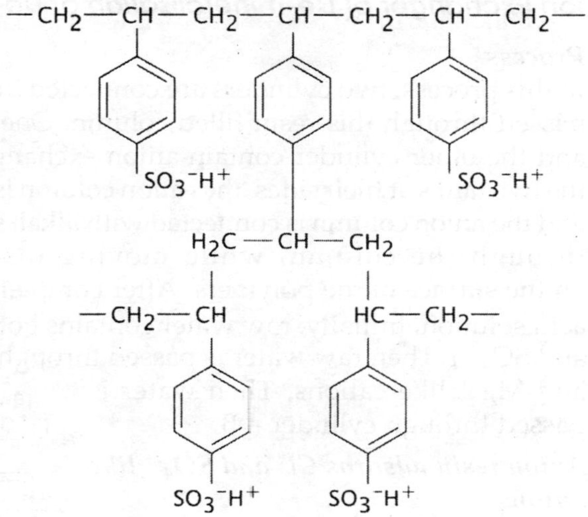

Anion Exchanger

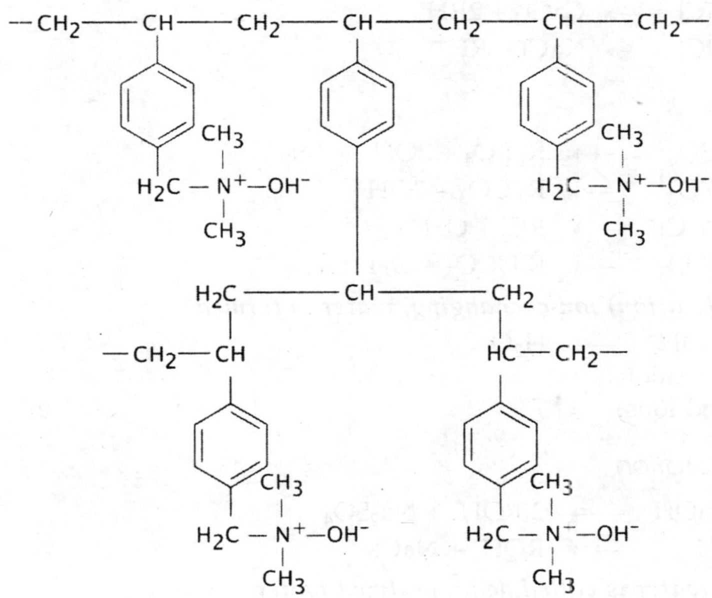

Ion Exchanger or De-mineralisation or De-ionisation

Process

In this process, two cylinders are connected in series. Using a circulation pump, water passed through this resin filled column. One cylinder contain cation exchange resin and the other cylinder contain anion exchange resin each column is connected with the two tanks at their sides, the cation column is connected with acid containing solution and the anion column is connected with alkali solution containing tank. Water is passed through the column, while moving inside the tank all ions are adsorbed on the surface of the polymers. After completion resins are regenerated by alkali and acid solution. Initially, raw water contains both ions (+ve = Ca^{2+} and Mg^{2+}), (–ve = Cl^- and SO_4^{2-}). Then raw water is passed through cylinder – A. Cation resin adsorbs Ca^{2+} and Mg^{2+} like cations. Then water is passed through cylinder – B

Anion resin adsorbs Cl^- and SO_4^{2-} like anions

Adsorption ion-exchange reactions:

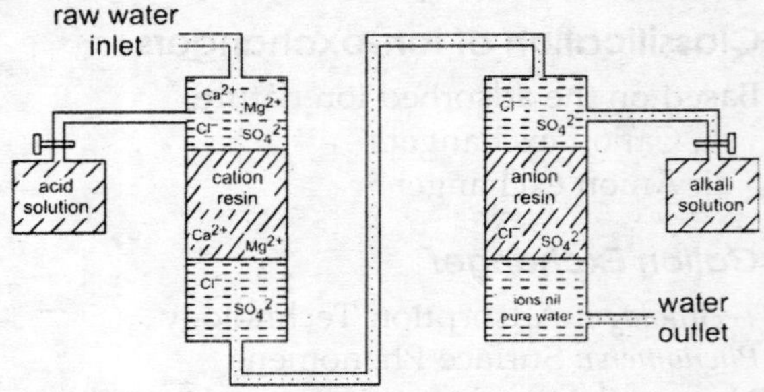

Fig. 3.27

Cation Reactions

$$2\,RH^+ + Ca^{2+} \longrightarrow R_2Ca + 2\,H^+$$
$$2\,RH^+ + Mg^{2+} \longrightarrow R_2Mg + 2\,H^+$$
$$RH^+ + Na^+ \longrightarrow RNa + H^+$$
$$RH^+ + K^+ \longrightarrow RK + H^+$$

Regeneration Reactions

$$R_2Ca + 2HCl \longrightarrow CaCl_2 + 2RH^+$$
$$RNa + HCl \longrightarrow NaCl + RH^+$$

Anion Reactions

$$2ROH^- + SO_4^{2-} \longrightarrow R_2\,SO_4 + 2OH^-$$
$$2ROH^- + CO_3^{2-} \longrightarrow R_2\,CO_3 + 2OH^-$$
$$ROH^- + Cl^- \longrightarrow RCl + OH^-$$
$$ROH^- + HCO_3^- \longrightarrow R\,HCO_3 + OH^-$$

While (neutralisation) ion exchanging, water is formed

$$H^+ + OH^- \longrightarrow H_2O$$

Cation Anion
(Liberated ions)

Anion Regeneration

$$R_2SO_4 + 2NaOH \longrightarrow 2\,ROH^- + Na_2SO_4$$
$$RCl + NaOH \longrightarrow ROH^- + NaCl$$

Final purified water is called demineralized water

3.21 CATALYSIS

Role of the adsorbent in catalysis or adsorption of contact theory

Reaction: **Reduction reaction of alkene**

Mechanism: Adsorption and Surface Phenomena

Reactant: Alkene ($CH_2 = CH_2$)

Catalyst: Nickel (Heterogeneous Catalyst)

Reaction

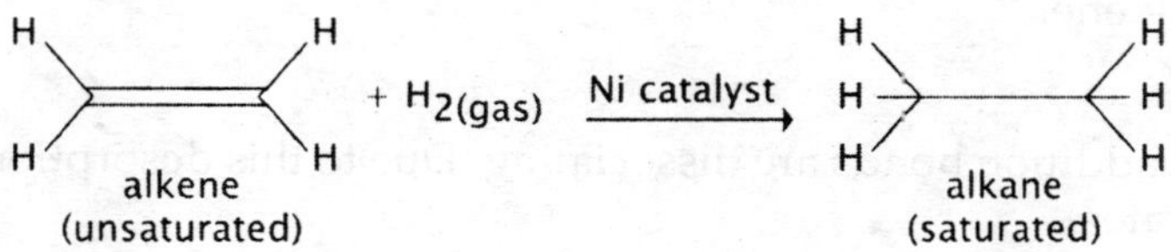

Mechanism and Steps

Step I

Surface complex formation

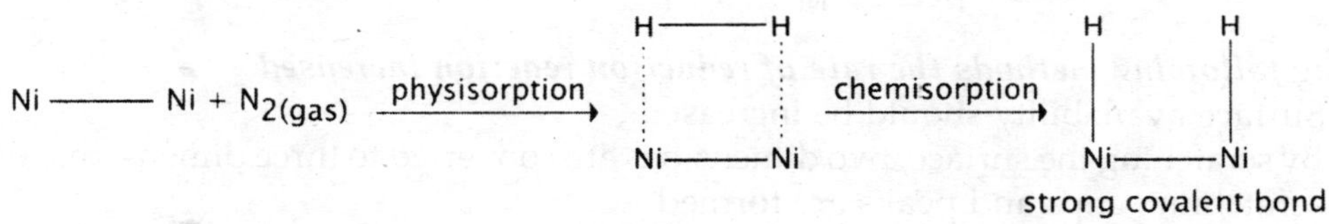

Initially, gaseous molecules are adsorbed on the catalyst surface. During adsorption, first physisorption occurs which is followed by chemisorption. Isolated active centres are adsorbing the gaseous molecule with emission of heat energy. That's why it is called an exothermic reaction.

Step II

In this step, an activated complex is formed with the Nickel surface. Activated unstable intermediate shelf life is very less.

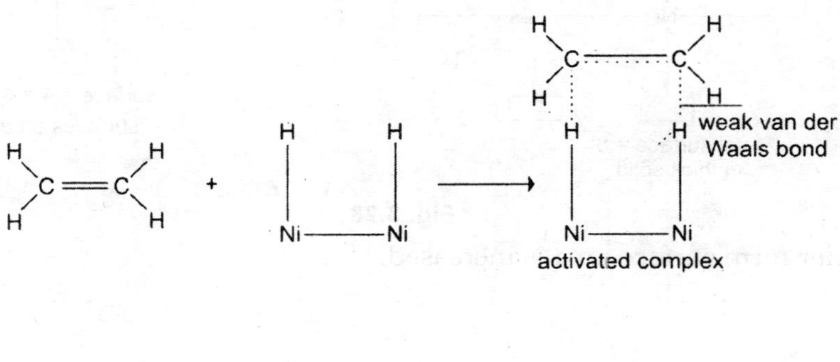

Step III

Decomposition

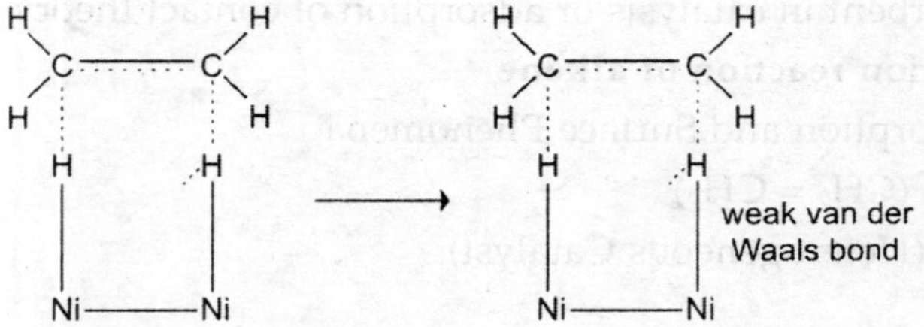

In this step, nickel bond is converted to weak van der Waals bond. Then hydrogen is added with alkene.

Step IV

After hydrogen addition bonds are dissociating. Due to this desorption nickel catalyst surface regenerates.

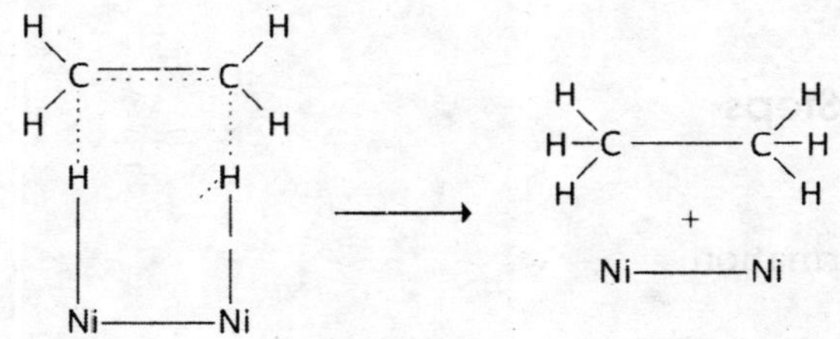

Using following methods the rate of reduction reaction increased

1. Surface availability should be increased
2. By scratching the surface, two dimensions are converted to three dimensions. While scratching, gaps and peaks are formed
3. By breaking the catalyst, surface area is increased. Maximum reduction reactions are carried out with powder catalysts.
4. Promoters are added to increase the rate of the adsorption.

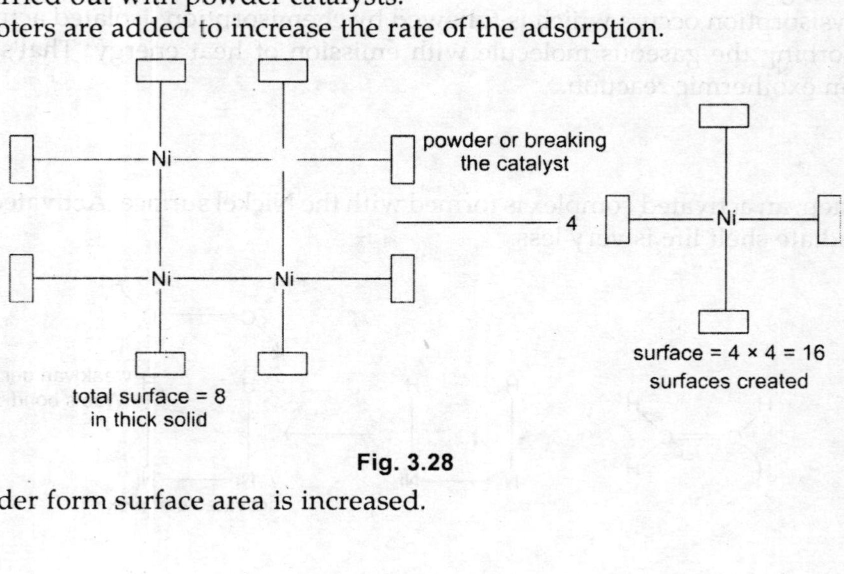

Fig. 3.28

In powder form surface area is increased.

Scratched Surface Two Dimensional Structures

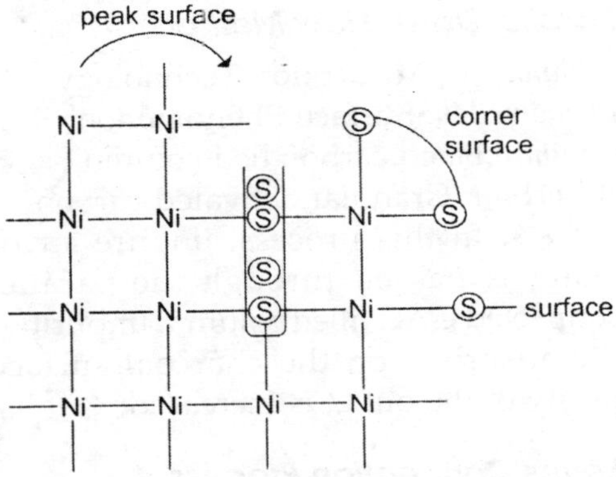

Fig. 3.29

Promoter's addition

Promoter is the material which increases the active surface of the catalyst and creates distance between two catalyst surfaces. Because of these promoters, rate of adsorption increases.

Role of promoters

- It increases the rate of reaction
- Promoter Changes (covalent bond) chemisorptions to physisorption (weak van der Waals bond)
- Desorption is very fast with catalyst and promoters.

Example, Pd/C - Palladium (catalyst)/Carbon (promoter)

Molybdenum acts as a promoter in Ni/H_2 reaction.

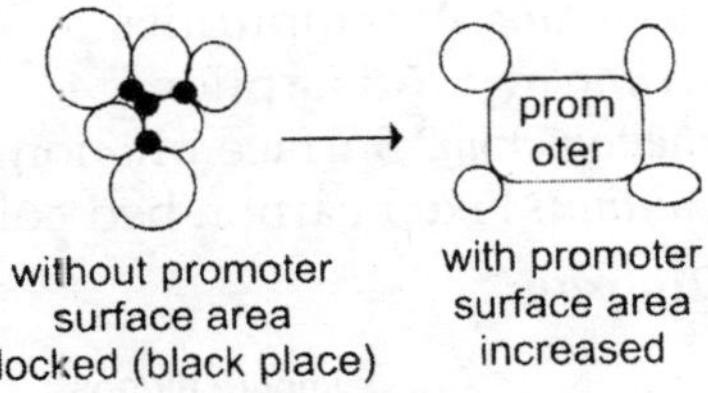

Fig. 3.30

Catalytic Poison

Catalytic chemical reactions which decrease the rate of adsorption and the catalytic surface are called catalytic poison. Compared to reactants and catalytic poison, catalytic poison adsorbs very fast

For examples, on platinum surface, carbon monoxide (CO) forms a thin layer. We have to use a specific catalyst for a specific reaction.

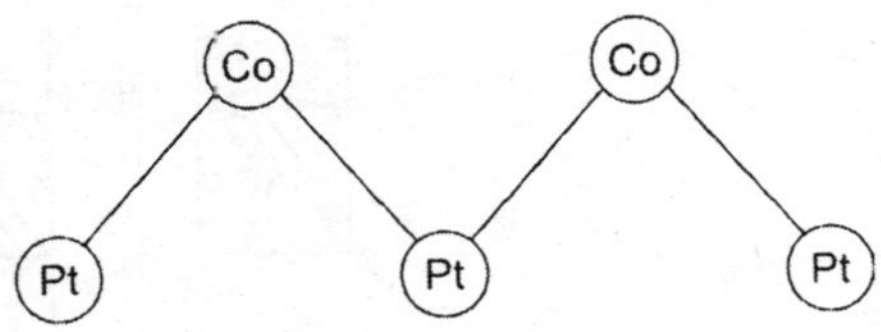

Fig. 3.31

Reason: To achieve the efficient product in catalytic chemical reactions. Specific catalyst is important.

3.22 ACTIVATED CARBON IN AIR PURIFICATION AND WATER PURIFICATION: POLLUTION REMOVAL

Reason for carbon as an adsorbent:

- Large surface
- Low cast

Types of Carbon: Based on Mesh size carbon is classified into two types:

A. Granular activated carbon ($10^4 - 10^8$ μ)

B. Powder activated carbon ($10^1 - 10^2$ μ)

Flow chart 3.1

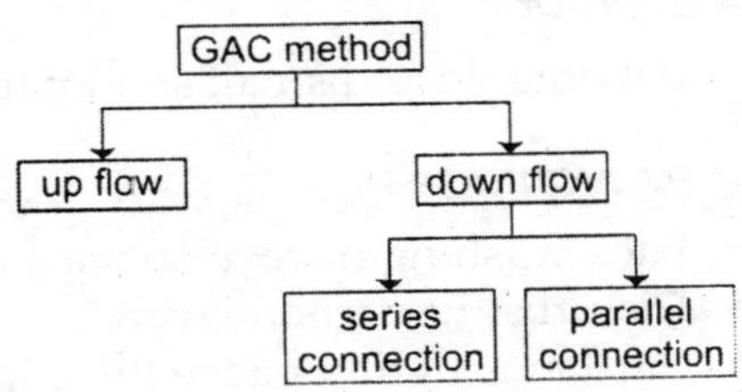

Using Granular Activated Carbon

Parallel Down Flow Method

Technology: Adsorption Technology
Phenomenon: Surface Phenomenon
Column: Fixed carbon bed column
Adsorbent: Granular activated carbon
Process: In this process, impure air or water is passed through the parallel connected GAC filled column. Impurities are adsorbed on the charcoal surface purity of the outlet is increased.

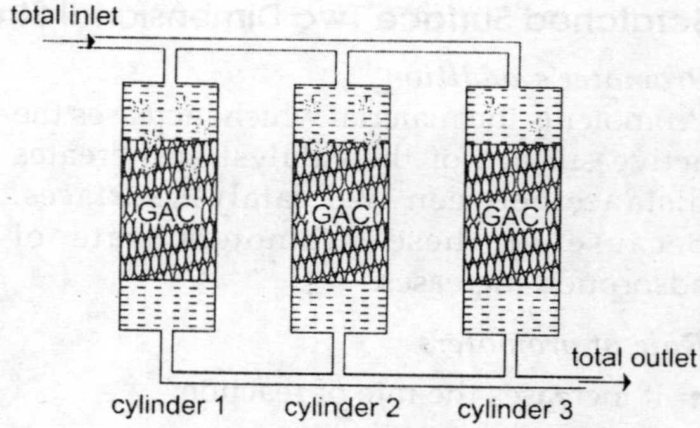

Fig. 3.32

Series Connection Process

Adsorbent: Granular Activated Charcoal
Adsorbate: Air impurities or Water impurities
Technology: Adsorption
Phenomenon: Surface phenomenon.
Column: Fixed carbon bed column
Diagram

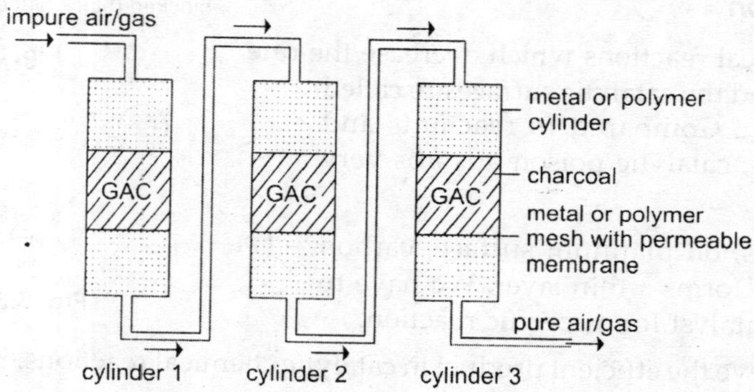

Fig. 3.33

Towards gravitational force impure gas/water is passed through carbon filled cylinders. Impurities adsorb on charcoal surface. Pure air or water come out. It is collected in tanks and cylinders.

Advantage

- All organic and small solid particles are adsorbed in a single step.

Disadvantages

- Back washing or reverse washing
- Disturbs the carbon layer.
- Carbon is tightly packed inside the column

Differences between series and parallel methods:

No.	Series	Paralled
1.	Polluted air is passed through more cylinders	Polluted air is passed through only one cylinder
2.	More surface area coverage	Less surface area coverage
3.	More efficient	less efficient

Up Flow Fixed Bed Conductor Method

Technology: Adsorption
Phenomenon: Surface Phenomenon
Column: Fixed Bed Column
Adsorbent: Granular Activated Charcoal
Adsorbate: Gas or Water impurities

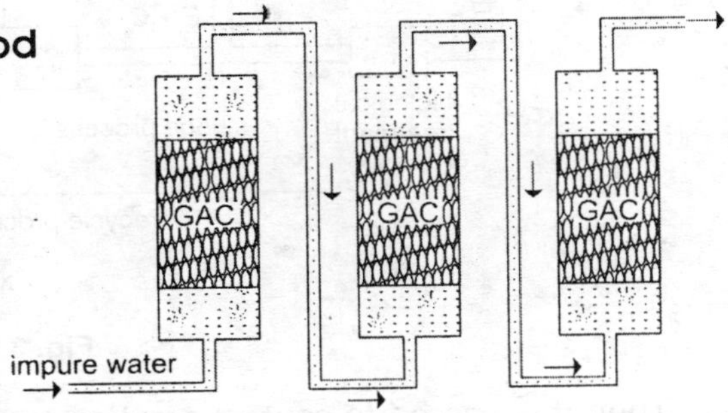

Fig. 3.34

Process

In this method, carbon is kept at the centre of metal or polymer Cylinders then impure air or polluted water is passed opposite to the gravitational force. Cylinders are connected in series, inside go the carbon adsorbing impurities and purified water comes out.

Advantage

- Most adsorbed materials have some density. These dense molecules try to come towards the gravitational force. So efficient purification occurs.

Disadvantages

In up-flow method, air passes through the layer during this time and length increases because of disturbed packed carbon, due to the circulation, carbon particles are moving and fine carbon escapes from the bed (Carbon is free flow).

Pollution Control by Powdered Activated Carbon or Industrial Effluent Purification

Technology: Adsorption Technology

Phenomenon: Surface phenomenon

Adsorbent: Powdered Charcoal

Adsorbate: Impurities

Process Structure:

a. Effluent contain colour, ions, organic molecules, toxic gases and inorganic salts
b. Effluent (subjected) is passed through biological treatment and Algae, Bacteria, Fungus are removed.
c. Charcoal is added to effluent and it adsorbs colour, organic molecules, inorganic, toxic gases on its surfaces.

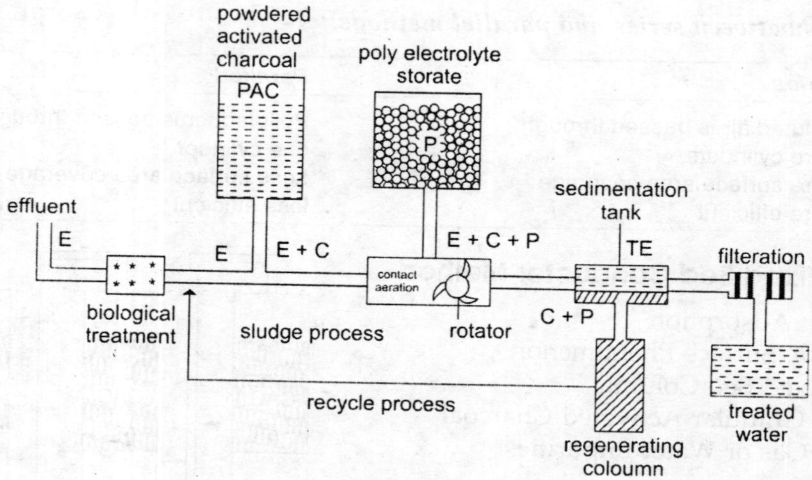

Fig. 3.35

d. Then it is passed to contact aeration tank. Oxygen is circulated by moving pipes to increase the COD and BOD.
e. Polyelectrolyte is added to remove the ions.
f. Then it is passed through the sedimentation tank and Charcoal and Polyelectrolyte settle from here
g. Then it is passed through the filtration column. Finally, treated water comes out.
h. Charcoal purified by regenerated column.

Applications of Charcoal

- Using charcoal, odour is controlled in industries.
- hydrogen sulphide is adsorbed by charcoal and pollution is controlled.
- To adsorb hydrocarbons from air and adsorb colour, charcoal is used.
- To clean sugar solution, vegetable oils, chemicals and pharmaceuticals products.
- To adsorb toxic gases in sipcot areas, it is packed in chemical masks.
- To adsorb vapour and sound in large auditoriums, charcoal mixed composites are used.
- To adsorb organic solvents and organic acids at low temperatures.
- To adsorb NH_3 from inorganic solids, charcoal treatment is given.

GLOSSARY

Adsorption: Gas or liquid concentration increases on the solid surface.
Absorption: Gas or liquid concentration increases throughout the body.
Solution: Solvent + Solute
Physisorption: Adsorption by weak van der Waals bond.
Chemisorption: Adsorption by strong covalent bond.
Adsorbent: Surface provider
Adsorbate: Surface seeker.

Adsorption isotherm: A graphical representation of pressure and adsorption.
Desorption: Concentration of gas or liquid decreases from the surface.
Catalyst: Which increases the rare of the reaction.

Assessment Questions

Answer all questions (2 marks)

1. Define adsorption and adsorbate.
2. What is sorption?
3. What is chemical adsorption or chemisorptions? Give an example.
4. What is physical adsorption or physisorption? Give an example.
5. Mention some important characteristics of adsorption.
6. How does chemisorption differ from physisorption?
7. Write any two differences between physisorption and chemisorption.
8. How will you increase the activity of an adsorbent?
9. Explain the effect of temperature on adsorption.
10. Explain the function of activated charcoal with example.
11. Define adsorption? What is an adsorption charcoal with example?
12. Define adsorption?
13. What is an adsorption isotherm?
14. What is Freundlich's adsorption isotherm?
15. What is Langmuir's adsorption isotherm?
16. Derive mathematically the expression of Langmuir's adsorption isotherm.
17. What is the demerit of Langmuir's adsorption isotherm?
18. What are promoters?
19. What is catalytic poisoning?
20. What is the effect of temperature and pressure on the adsorption of hydrogen gas on charcoal?
21. How arsenic poisoning is removed from the body?
22. Define ion-exchange adsorption?
23. How is evaporation of water in lakes minimized?
24. Explain promoters?
25. What is physical adsorption?
26. What is chemical adsorption?
27. Explain differences between chemical and physical adsorption?
28. Explain sorption?
29. Explain adsorption of demineralization of water.
30. Explain adsorption in catalytic reaction.
31. What are the conditions of Freundlich's adsorption isotherms?
32. Explain adsorption.
33. Explain adsorption isotherm.
34. Define Freundlich's adsorption isotherm.
35. Explain Langmuir's adsorption isotherm.
36. Define promoters.

37. Explain adsorbate.
38. Explain sorption.
39. Define catalytic poisoning.
40. Explain adsorbent.
41. Explain types of adsorption.
42. What is covalent bond?
43. What is ionic bond?
44. Explain characteristics of adsorption.
45. Explain types of adsorption isotherms?
46. Explain limitations of Freundlich's adsorption isotherm?
47. What are the disadvantages of Freundlich's adsorption isotherms?
48. Derivation of Freundlich's adsorption isotherms?
49. Explain GAC.
50. Explain PAC.

Answer all questions (16 marks)

1. Distinguish between physical adsorption and chemical adsorption.
2. Derive an expression for Langmuir unimolecular adsorption isotherm. What are the limitations?
3. Explain adsorption theory or contact theory with examples.
4. Explain the role of adsorption in catalytic reaction.
5. Explain the classification and function of ion-exchangers.
6. Define the term adsorption and list of applications.
7. Explain the role of adsorption in demineralization of water.
8. Give any three factors on which adsorption depends
9. Derive Freundlich's adsorption isotherm. Give the conditions in which it fails.
10. Discuss the factors which influence adsorptions of a gas on a solid.
11. Derive an expression for Langmuir's adsorption isotherm. Show that at normal pressure Langmuir's adsorption isotherm becomes identical with Freundlich's adsorption isotherm.
12. Define the terms adsorbent and adsorbate giving suitable examples.
13. Mention any three factors which influence adsorption of a gas on a solid.
14. Derive Langmuir's absorption isotherm. Write the factors on which adsorption depends.
15. Derive the Langmuir's absorption isotherm and interpret the results at
    a. Low pressure and
    b. High pressure
16. Define adsorption isotherm. Explain the various types of adsorption isotherms.
17. Explain the role of adsorbents in pollution abatement.
18. i. Derive Langmuir's adsorption isotherm.
    ii. Explain the factors influencing adsorptions of solutes from solutions.
19. What are the roles of adsorbents in pollution abatement?
20. i. Derive the Freundlich's adsorption isotherm.
    ii. Explain the differences between physisorption and chemisorption.
21. Write notes on the role of adsorbents in catalysis.

4

Non-Conventional Energy Sources and Storage Devices

4.1 INTRODUCTION

Nowadays energy conversion and saving energy is an important role. Because of population growth, need of energy increases.

Why is this topic coming in engineering curricula?

Engineering students should think about the conversion methods of energy by different ways. Electrical students should think about the energy saving methods and how to fulfill our energy requirement. Dynamos are already developed. Electronics students must think about the devices which can convert other energies into useful energies. For example, while walking, mechanical energy is wasted. We can convert this energy into electrical energy. Based on this, pressing Torches (batteries) are developed solar batteries are developed.

Energy conversion machines should be designed by mechanical students 100% energy conversion engines should be designed. Because of these types of developments, students should read about the energy sources. Nowadays small kids are also thinking about energy conversion. Small kids are thinking about the thunder energy conversion. So try to participate in energy conversion, save energy and save natural resources.

4.2 ABOUT THE ENERGY SOURCES

- *Fossil fuels:* Naturally biodegraded plant and animal products are called Fossil fuels, for example, Coal, Petroleum and Natural gas. These fuels are called non-conventional energy sources. Fossil fuels are already converted as energy source. so it is not necessary to convert them again.

In fossil fuels chemical energy was stored. This stored chemical energy is converted to heat energy and used in many ways. This chemical energy is formed by rearrangement of outer (valence) electrons. Nowadays these energy sources availability on earth is decreasing. So, we have to save the energy source, and look up for the new energy source.

The energy sources are classified into three types

Flow chart 4. 1

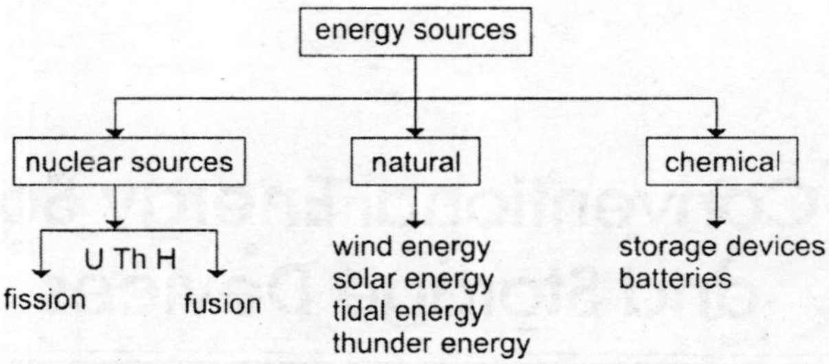

Nuclear Energy or Radioactivity

Important notes to write nuclear reaction

Name	***Symbol***	***Atomic number***
Actinium	Ac	89
Thorium	Th	90
Protoactenium	Pa	91
Uranium	U	92
Neptunium	Np	93
Plutonium	Pu	94
Polonium	Po	84
Lead	Pb	82
Rubidium	Rb	37

Isotopes of hydrogen $_1H^1$ (Protium), $_1H^2$ (Dutrium) and $_1H^3$ (tritium)

Important Radioactive Particles

α	Alpha	$_2He^4$	+ve charge
β	Beta	$_{-1}e^0$	–ve charage
γ	Gama	$_0\gamma^0$	Neutral
n	Neutron	$_0n^1$	Neutral
ρ	Positron	$_{+1}e^0$	+ve charge

Atomic symbol ($_AX^Z$)

X - Atomic symbol, Z - Atomic number and A - mass number.

Energy from natural resources

Radioactivity

Emission of powerful radiation continuously by the deformation of an element is called radioactivity.

Example of Radioactive metals—Uranium (U) and Thorium (Th)

4.3 RADIOACTIVE DECAY

The nucleus of an atom emit radiation spontaneously, (N) because of this continuous radiation process one parent atom changes into a new atom. These changes are called radioactive decay. Nucleus may emit any (n and positron) particle.

Example:

$${}_{92}U^{238} \longrightarrow {}_{90}U^{234} + {}_{2}He^{4}$$

4.4 TYPES OF RADIOACTIVE DECAY

Nuclear reactions follow the law of mass action.

Mass and Atomic number of right hand side = Left hand side

Type I (α - Particle emission reaction (${}_{2}He^{4}$)

Example,

$${}_{92}U^{238} \longrightarrow {}_{90}U^{234} + {}_{2}He^{4}$$

$${}_{84}Po^{215} \longrightarrow {}_{82}Pb^{211} + {}_{2}He^{4}$$

Atomic number decreases by 2 and Mass number decreases by 4 in this decay.

So the product is ${}_{A-2}X^{Z-4}$

Type II (β - Particle emission reaction (${}_{-1}e^{0}$)

Example,

$${}_{90}Th^{*233} \longrightarrow {}_{91}Pa^{233} + {}_{-1}e^{0}$$

So the product is ${}_{A+1}X^{Z}$

In this decay process atomic number increases by 1

Type III (γ - Particle emission reaction)

Example,

$${}_{3}Li^{7} + {}_{1}H^{1} \longrightarrow {}_{4}Be^{*8} \longrightarrow {}_{4}Be^{8} + \gamma$$

$${}_{7}N^{14} + {}_{1}H^{1} \longrightarrow {}_{8}O^{*15} \longrightarrow {}_{8}O^{15} + \gamma$$

Sometimes, higher amount of energy released from the type of reactions.

Type IV Positron particle emission reactions (${}_{+1}e^{0}$)

Example,

$${}_{1}H^{1} \longrightarrow {}_{0}n^{1} + {}_{1}e^{0}$$

During the emission of positron, product Atomic number decreases by 1.

Type V Neutron particle emission reaction (${}_{0}n^{1}$)

Example,

$${}_{1}H^{1} + {}_{1}e^{0} \longrightarrow {}_{0}n^{1}$$

$${}_{6}C^{12} + {}_{1}H^{2} \longrightarrow {}_{7}N^{*14} \longrightarrow {}_{7}N^{13} + {}_{0}n^{1}$$

$${}_{8}O^{16} + {}_{1}H^{2} \longrightarrow {}_{9}F^{*18} \longrightarrow {}_{9}F^{17} + {}_{0}n^{1}$$

These neutrons are used to attack the Uranium to produce higher amount of energy.

4.5 DIFFERENCE BETWEEN NUCLEAR REACTIONS AND CHEMICAL REACTIONS

No.	*Nuclear reactions*	*Chemical reactions*
1.	Reactions based on nucleus composition Nucleus = Proton + neutron	Reactions based on valence electron (last orbital electrons)
	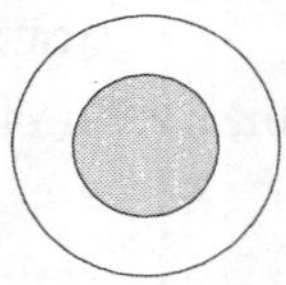	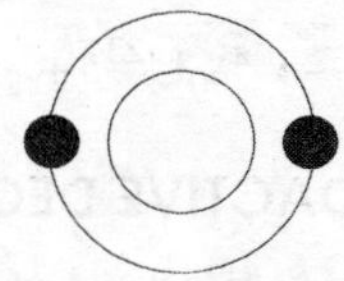
2.	New element formed	New molecule formed
3.	Reactions release maximum higher amount of energy (exothermic reaction)	Reactions release or absorb energy (exothermic or endothermic reaction)
4.	Rate of nuclear reactions does not depend upon physical properties. Pressure (P), Temperature (T), Concentration (C)	Rate of reaction depends upon Pressure, Temperature and Concentration
5.	α, β, and γ particles are emitted	No particle emission

4.6 TYPE OF NUCLEAR REACTIONS

The process to change the target atomic nucleus by attacking light atoms or nuclei is called nuclear reaction.

Based on the nature of the reaction, nuclear reactions are classified as follows:

i. Nuclear fission reactions
ii. Nuclear fusion reactions
iii. Projectile capture reactions or adsorbed new nuclei
iv. Capture and emission reactions or Particle adsorbed and emitted reactions

Types of Nucleus

Based on the reactant and product nature, a nucleus can be classified into two types:

- Projectile or Bombarded or Insisting nucleus
- Emitted nucleus

Type I Bombarded particle capture reaction

In this reaction the bombarded particle is added with the reactant nucleus and form a new atom with the emission of Gama.

Example,

$$_{37}Rb^{85} + {_0n^1} \longrightarrow {_{37}Rb^{86}} + \gamma$$

$$_{6}C^{12} + {_1H^1} \longrightarrow {_7N^{13}} + \gamma$$

Type II (Projectile capture reactions)

$$_{7}N^{14} + {_0n^1} \longrightarrow {_6C^{12}} + {_1H^3}$$

$$_{9}F^{19} + {_1H^1} \longrightarrow {_8O^{16}} + {_2He^4}$$

$$_{5}B^{11} + {_1H^1} \longrightarrow {_6C^{*12}} + {_2He^4} + {_4Be^8}$$

When Captured nucleus is added to the reactant nucleus, a new nucleus is formed and particles emitted.

4.7 NUCLEAR FUSION REACTION

Type III

Definition

The reaction where in lighter nuclei fuse together and form heavier nuclei emitting huge amount of energy is called the fusion reaction.

Example:
Solar energy, universe energy and star energy

Properties of Fusion Reactions

Fusion reactions occurs at higher temperatures. That's why it is called thermonuclear reactions. The Reaction temperature is more than 108° K (Kelvin). Due to this reaction higher amount of energy is released spontaneously.

Examples for fusion reactions

$${}_1H^2 + {}_1H^2 \longrightarrow {}_2He^4 + \text{Energy (3. 26 MeV)}$$
$${}_1H^2 + {}_1H^3 \longrightarrow {}_2He^4 + {}_0n^1 + \text{Energy (17. 6 MeV)}$$
$$4\,{}_1H^1 \longrightarrow {}_2He^4 + 2\,{}_1e^0 + \text{Energy (24 MeV)}$$
$${}_1H^1 + {}_1H^3 \longrightarrow {}_2He^4 + \text{Energy}$$

These reactions the occur in the universe. stars energy = 0. 02871 U = 26. 7 MeV and Sun = 3. 8 × 10^{20} mJ. In this reaction Mass loss is converted to energy.

***Type IV* Nuclear Fission Reaction**

The reaction wherein target nucleus is attacked by thermal neutrons and split the heavier nuclei into smaller nucleus with emission of particles and huge amount of energy is called nuclear fission reaction.

Example,

$${}_{92}U^{235} + {}_0n^1 \longrightarrow {}_{92}U^{236*} \longrightarrow {}_{56}Ba^{136} + {}_{36}Kr^{94} + 3\,{}_0n^1 + \text{Energy}$$

Types of Neutrons

Based on the velocity and energy of the neutron, it is classified into two types:

i. Thermal neutron (velocity, energy lower)
ii. Fast neutron (velocity, energy higher)

Other Nuclear Fission Reactions

$${}_{92}U^{231} + {}_0n^1 \longrightarrow [{}_{92}U^{235}]^* \longrightarrow {}_{54}Xe^{144} + {}_{38}Sr^{90} + 2{}_0n^1$$
$${}_{92}U^{231} + {}_0n^1 \longrightarrow [{}_{92}U^{235}]^* \longrightarrow {}_{55}Cs^{144} + {}_{37}Rb^{90} + 2{}_0n^1$$
$${}_{92}U^{231} + {}_0n^1 \longrightarrow [{}_{92}U^{235}]^* \longrightarrow {}_{56}Ba^{140} + {}_{36}Kr^{93} + 3{}_0n^1$$

(Unstable)

Diagram: Mechanism of the fission reactions
Energy released is based on the mass defect.

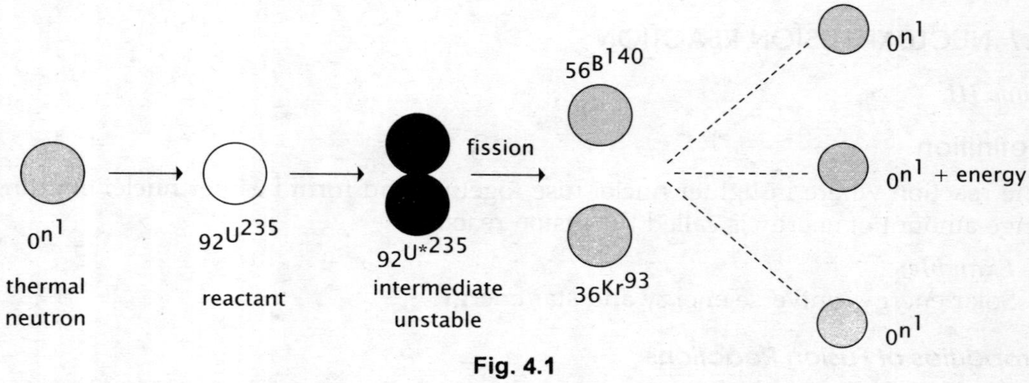

Fig. 4.1

Properties of Nuclear fission reactions

- Thermal neutrons are used to bombard the target nuclei.
- Heavy nucleus is used for fission reaction.
  Example, ${}_{92}U^{235}$ and ${}_{94}Pu^{239}$
- All the fission products are radioactive which emit radioactive particles (α, β and γ, etc).
- These reactions release more than one neutron.
- Huge amount of energy is released by these reactions based on the small mass loss.
- Nuclear fission reactions are chain reactions or self propagating reactions.
- Nuclear fission reactions can be controlled by adsorbing excess neutron.
- Atomic weights of fission products range from 72 to 162.
- Formed neutrons propagate the reaction. Number of neutrons formed from the first reaction is called multiplication factor.
- If the reaction releases 2 neutrons. 2^n is multiplication factor.

4.8 NUCLEAR CHAIN REACTIONS

The self propagating and fast neutron producing reactions are called nuclear chain reactions. These Neutrons attack the target nuclei again and again producing higher amount of energy. Uranium reaction producing three neutrons in each step. Multiplication factor is 3^n

Types of chain reaction: Controlled chain reaction, uncontrolled chain reaction

Schematic structure of uncontrolled chain Reaction
Formed neutrons or involved Uranium atom = $3^1 + 3^2 + 3^3 + ... 3^n$

In nuclear fission reactions, all emitted neutrons continuously react with uranium and produce huge amount of energy. This is called uncontrolled fission reaction.

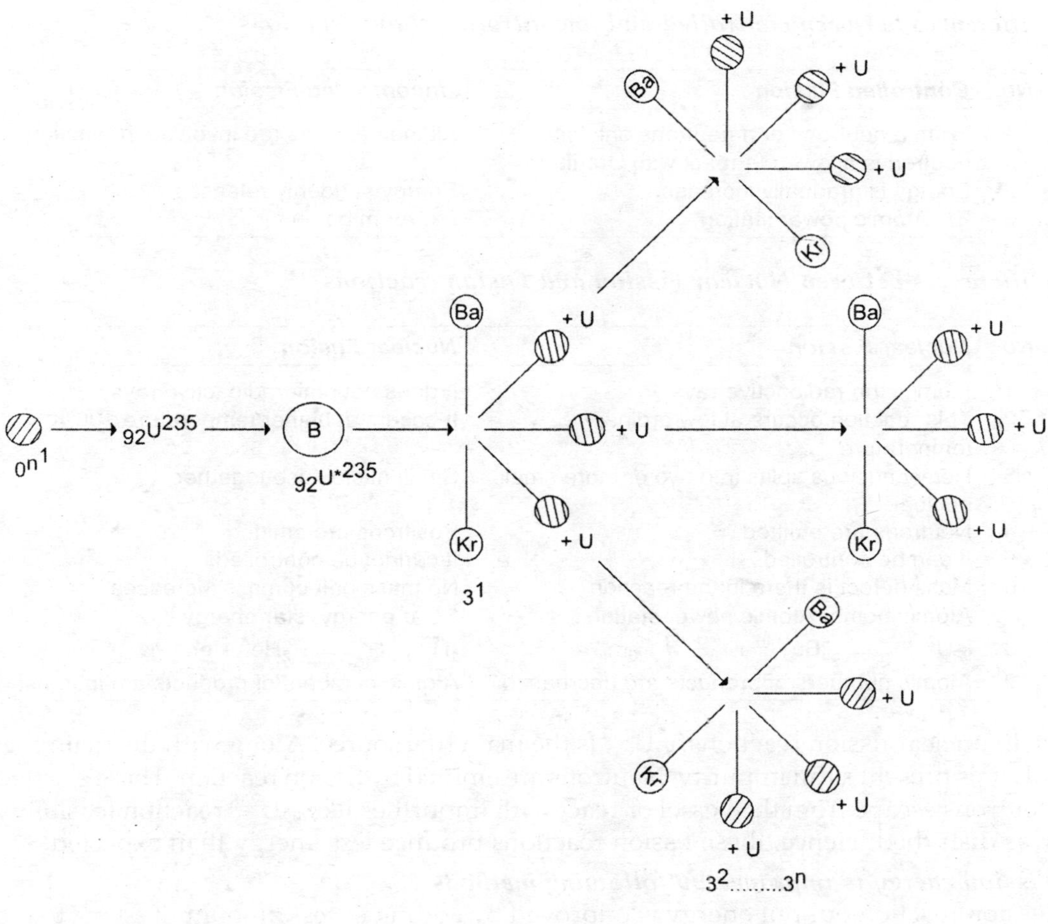

Fig. 4.2

Controlled Chain Reaction

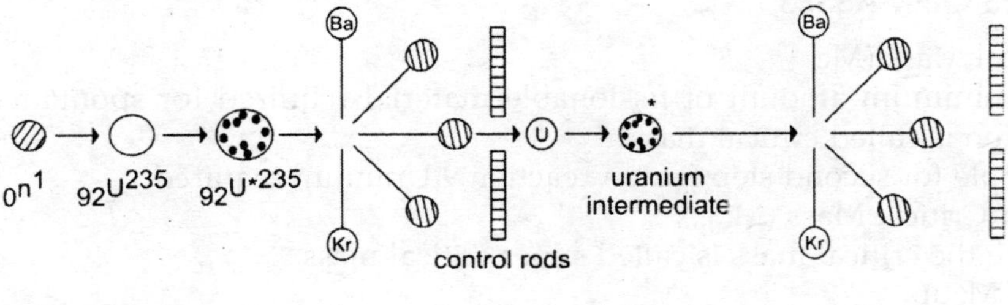

Fig. 4.3

Differences between controlled and uncontrolled chain reactions

No.	Controlled Fission	Uncontrolled Fission
1.	From 3 neutrons or n neutrons only one neutron is allowed to react with Uranium	All the neutrons are involved in reaction
2.	Energy is gradually increased	Energy suddenly releases
3.	Ex: Atomic power station	Ex: Atom bomb

Differences between Nuclear Fission and Fusion reactions

No.	Nuclear Fission	Nuclear Fusion
1.	It emits the radioactive rays	It does not emit radioactive rays
2.	This reaction occurs at low ordinary temperature	It occurs at higher temperature 106°K
3.	Heavy nucleus splits into two or more small nuclear	Small nuclei fuse together
4.	Neutrons are emitted	Positrons are emitted
5.	It can be controlled	It cannot be controlled
6.	Mass defect is there in the reaction	No mass defect mass increases
7.	Atomic bomb, Atomic power station	Solar energy, star energy
8.	${}_{92}U^{236} \longrightarrow {}_{56}Ba^{140} + {}_{36}Kr^{93} + {}_{0}n^{3}$	${}_{1}H^{2} + {}_{1}H^{2} \longrightarrow {}_{2}He^{4}$ + energy
9.	Atomic numbers of products are decreased	Atomic numbers of products are increased

In nuclear fission reactions, ${}_{92}U^{235}$ is the main fuel source. Along with the main fuel ${}_{92}U^{238}$ is present as an impurity. Neutrons are emitted by fission reaction. These emitted neutrons escape from the vessel or react with impurities like ${}_{92}U^{238}$ reaction feasibility was disturbed. Hence, these fission reactions produce less energy than expected.

Fission energy is improved by following methods:

Fission reaction out put energy is improved by adding excess amount of ${}_{92}U^{235}$ to the reactor.

Because of this, excess supplied neutrons do not escape from the reactor. Thermal neutrons should be supplied continuously.

Example, Ist step – 3 Uranium (3 + 2) excess
IInd step – 9 Uranium (9 + 3) excess

4.9 TYPES OF MASSES

a. Critical Mass (Mc)
The minimum amount of fissionable material required for spontaneous chain reaction is called critical mass.
Example for second step fission reaction 9 Uranium required

b. Super Critical Mass (M_{sup})
Above the critical mass is called super critical mass.
Mc < Msup

c. Sub Critical Mass (M_{sub})

Below the critical mass is called subcritical mass.

$Mc > M_{sub}$

4.10 MASS DEFECT

The mass different between nuclear reactant and nuclear product is called mass defect.

Example,

$${}_{92}U^{235} \longrightarrow {}_{56}Ba^{138} + {}_{36}Kr^{93} + 3\ {}_{0}n^{1}$$

Total mass = 236 Total Mass = 138 + 93 + 3 = 234

Mass Defect = Mr – Mp = 236 – 234 = 2

The above mass defect converted to energy as per Einstein equation.

$E = mc^2$

4.11 NUCLEAR ENERGY

The energy released by either fission or fusion reaction is called nuclear energy.

1. If the energy is produced by nuclear fission, it is called nuclear fission energy
2. If the energy is produced by fusion reaction, it is called nuclear fusion energy

Heavy Isotopes: ${}_{92}U^{235}$ and ${}_{94}Pu^{239}$

Einstein equation

$E = mc^2$

E – Energy, M – Mass (mass defect), C – Velocity of light

Problem based on nuclear energy

Energy equation is E = mc^2

Mass of neutron (n) = 1. 009 amu

Velocity of light (C) = 3 × 108 m/s

amu – Atomic mass unit = 1. 66 × 10–27 kg

1ev = 1. 6022 × 10–19 J

Problem

$${}_{92}U^{235} + {}_{0}n^{1} \longrightarrow {}_{56}Ba^{141} + {}_{36}Kr^{92} + 3\ {}_{0}n^{1}$$

Reactant Product

Mass Difference

Reactant

Reactant Mass ${}_{92}U^{235}$ = 235. 120

Neutron Mass ${}_{0}n^{1}$ = 1. 0098

Total mass = 236. 129 ... 1

Product

Mass of Barium = 140. 910

Mass of Krypton = 91. 910

Mass of 3 Neutron = 3. 027 = 3 × 1. 009

Total Mass = 235. 847 ... 2

$$\text{Mass Defect} = \text{Equation (1)} - (2)$$
$$= 236.129 - 235.847$$
$$M = 0.282 \text{ amu} \quad \ldots 3$$

Substitute *m* value in Einstein equation

$$E = mc^2 \quad \ldots 4$$
$$1 \text{ amu} = 1.66 \times 10^{-27} \text{ kg} \quad \ldots 5$$

Substitute (3), (5) in equation (4)

$$E = 0.282 \times 1.66 \times 10^{-27} \text{ kg} \times (3 \times 10^8) \text{ kg m}^2 \text{ s}^{-2}$$
$$= 0.282 \times 1.66 \times 9 \times 10^{-27} \times 10^{16} \text{ kg m}^2 \text{ s}^{-2}$$
$$= 4.21 \times 10^{-11} \text{ kg m}^2 \text{ s}^{-2}$$
$$= 4.21 \times 10^{-11} \text{ Joule} \qquad \text{Where } J = \text{kg m}^2 \text{ s}^{-2}$$

Energy in electron volt
Joule value divided by eV value

$$E = \frac{4.21 \times 10^{-11} J}{1.6022 \times 10^{9} J/eV}$$

Released energy (E) = 2.62×10^8 eV
For 1 amu = 931 MeV is energy

Nuclear Reactor

Nuclear fission reaction is carried out under a controlled manner in an arranged equipment. The heat energy is converted into useful electrical energy.

The arranged setup or device where the fission reaction is carried out under a controlled manner is called **nuclear reactor or pile.**

Pile: where the lot of material is inserted.

Use of nuclear reactor: Heat energy is converted to electrical energy

Components of Nuclear Reactor

1. Reactor core
2. Fuel rods
3. Moderators
4. Control rods
5. Coolants
6. Pressure vessel
7. Protective shield
8. Turbine

Reactor Core

The setup which consists of fuel rods, moderator, coolant and control rods is called reactor core.

In reactor core, the fission reaction is controlled.

Fuel Rods

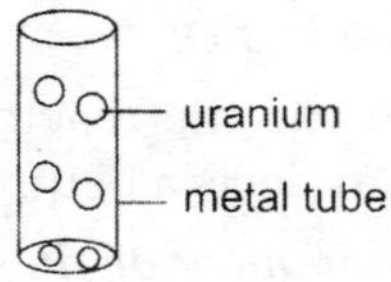

Fig. 4.4

The fissionable material $_{92}U^{235}$ is the fuel of nuclear reactor. Enriched $_{92}U^{235}$ Uranium is filled in metal tubes and kept inside the reactor. Fuels react with thermal neutrons and produce heat energy by chain reaction.

Fuel – $_{92}U^{235}$

Control Rods

This rod controlls the rate of fission reaction and absorbs excess neutrons. It is movable up and down. It is kept in between fuel rods.

Upward — Rate of chain reaction is fast
Downward — Rate of chain reaction is slow

Control rods are made up off Cadmium (Cd), Boron (B)

Adsorption of Neutrons

$$_{92}Cd^{235} + {}_0n^1 \longrightarrow {}_{43}Cd^{114} + \gamma\text{-rays}$$
$$_5B^{10} + {}_0n^1 \longrightarrow {}_5B^{11} + \gamma\text{-rays}$$

Moderators

Moderators are also used to control the fission reaction. these reduce the kinetic energy of formed neutrons. Fission reactions produce fast neutrons.

Example for moderators: Heavy water (D_2O), light water (H_2O), Graphite rod and Berylium rod.

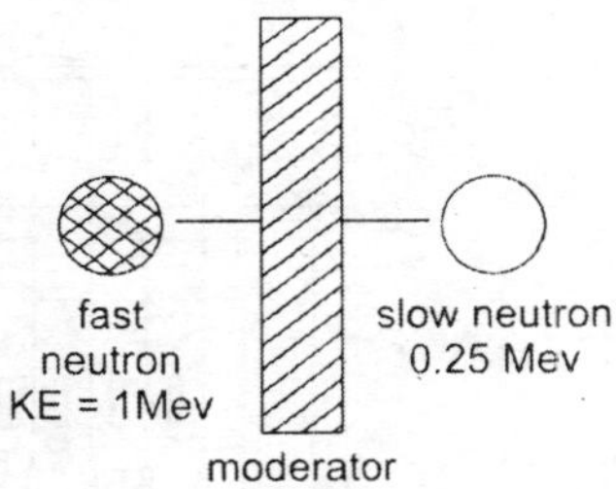

Fig. 4.5

Coolants

Fission reaction produces huge amount of heat energy. Due to this heat energy, high pressure is developed inside the nuclear reactor. The material used to reduce the heat energy inside the reactor by absorbing is called coolant. Heat energy absorbed by the coolant is used to produce steam. Coolant is circulated outside the nuclear reactor core.

Example, Water (light and heavy water), liquid metal (Na and K) and Air (CO_2)

Pressure Vessel

This maintains pressure inside the reactor core. Inside the reactor core 200 kg cm^{-2} pressure maintained by pressure vessal. It is along with coolant circulation.

Protective Shield

The total reactor core is covered by 10 meters and more than 10 meters thickness of concrete wall. It protect from radioactive rays emission.

Turbine

It rotates the wheel. Because of steam this wheel rotates and mechanical energy gets converted to electricity. It acts as a dynamo

Reflector

Inside the reactor core, reflecting material is present. Moderator acts as a reflector. It reflects a neutron inside the core. It does not allow neutrons to escape from the vessel.

Summary of the Reactor

Fuel Rod	–	$_{92}U^{235}$, $_{94}Pu^{239}$
Moderator	–	Heavy water (D_2O), light water (H_2O), Graphite and Beryllium
Control Rod	–	Cadmium (Cd) and Boron (B)
Coolant	–	Water, Liquid metal (Na, K) and Air (CO_2)
Pressure vessel	–	200 kg cm^{-2}
Protective shield	–	Concrete (10-meter thickness)

4.12 NUCLEAR REACTOR OR PILE

A nuclear power plant is used to prepare electricity from nuclear fission reactions. In a nuclear power plant, cooling purpose water is circulated inside. Based on this water, nuclear reactors are classified into two types:

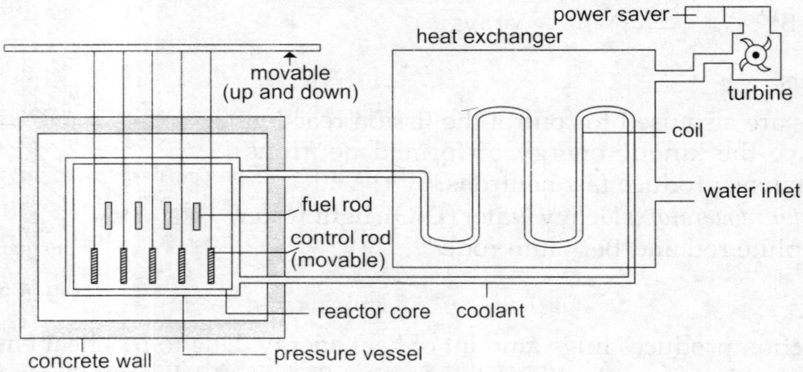

Fig. 4.6

Flow chart 4.2

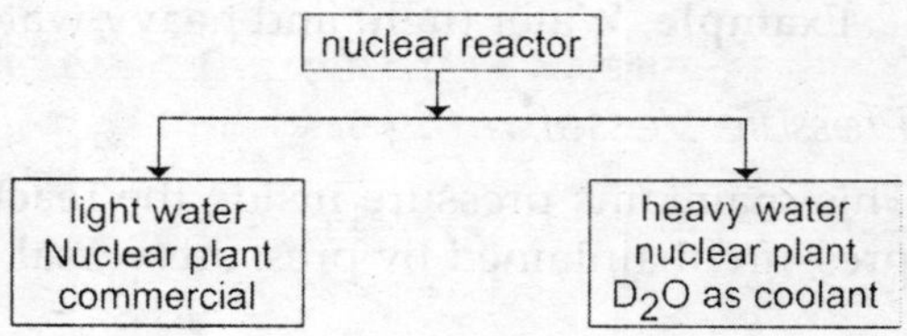

1. *Light water nuclear power plant:* If ordinary water is used for cooling purpose in a nuclear power plant, that is called light water nuclear power plant (H_2O is a coolant)
2. *Heavy water nuclear power plant:* If deutrated water is used as coolant, that is called heavy water nuclear power plant (D_2O)

Light Water Reactor Operating Procedure

Thermal neutrons are passed inside the nuclear reactor core. These neutrons bombard fuel rod [$_{92}U^{235}$] emitting high heat energy with nuclear fission products. Chain reaction can be controlled by moving the control rod (Cd) up and down. Light water is circulated inside the reactor and heat energy is absorbed by the coolant (H_2O). Hot water is passed through the heat exchanger. Heat exchanger contains seawater. Seawater absorbs heat and becomes steam. This steam operates the turbine, rotating. It is producing electricity, and it was saved by transformers. Nuclear wastes are collected in concrete vessels.

4.13 BREEDER REACTOR

Good nuclear fuel should be easily fissionable. ${}_{92}U^{235}$ and ${}_{94}Pu^{239}$ are the example for fissionable material. Based on the rate of fission reaction nuclides are classified into two types:

- Fissile Nuclides
- Fertile Nuclides

Fissile Nuclides

The nuclide that undergoes fission reaction readily is called fissile nuclides.

Example, ${}_{92}U^{235}$, ${}_{94}Pu^{239}$

Fertile Nuclides

The nuclide that does not undergo fission reaction readily is called fertile nuclides.

Example,

${}_{92}U^{238}$, Th^{232}

Definition for Breeder Reactor

The nuclear reactor in which fertile nuclides are converted to fissile nuclides is called breeder reactor. It breeds (spread or emit) more fissionable material.

Simplified

$$\text{Non-fissionable fuel} \xrightarrow{\text{breeds}} \text{more fissionable fuel}$$

$${}_{92}U^{238} + {}_{0}n^{1} \longrightarrow {}_{94}Pu^{239} \xrightarrow[2.436 \times 10^{4}]{\alpha} {}_{92}U^{235}$$

Diagrammatic Representation

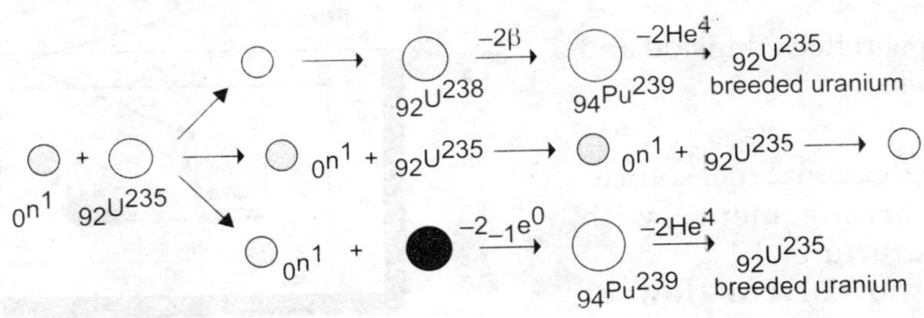

Fig. 4.7

The fuel used to produce thermal neutron is called the primary fuel. The fuel formed from (fertile) non-fissionable material is called the secondary fuel.

Energy Conversion From Natural Energy Sources

- Solar energy
- Wind energy

4.14 SOLAR ENERGY

In sun, huge amount of heat energy is produced by nuclear fusion reaction. This heat energy is emitted as radiation to earth.

This energy conversion is classified into two types:

1. Thermal conversion
2. Photo conversion

Flow chart 4.3

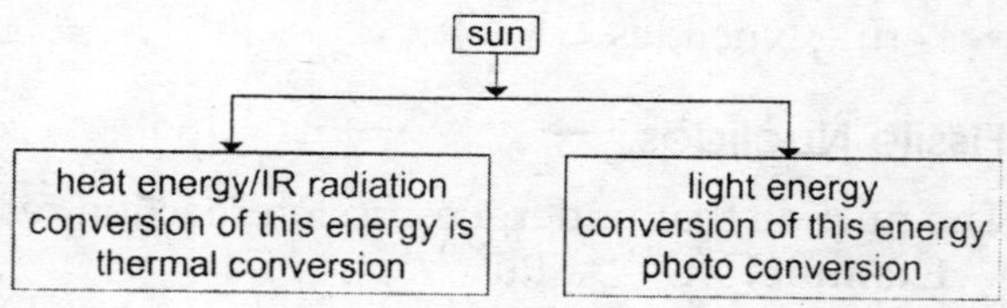

Thermal conversion

From sun, (heat energy) IR radiation energy conversion is called thermal conversion. Directly IR radiation is adsorbed as thermal energy.

Example:

Solar cooker and Solar water heater

Solar cooker

Instrument	–	solar cooker
Energy source	–	IR radiation of sun
Conversion	–	Thermal conversion
Principle used	–	Black body radiation and reflector

- Solar cooker contains a black coated metallic box inside.
- The box contains holes on the surface to keep the food vessel. Outside of the box is coated black.
- One end is connected with plane mirror to reflect the radiation.
- This whole system is covered by a glass plate.
- The cooker is kept towards sunlight.
- Radiation is reflected inside and food is boiled.

Advantages

- No need of external fuel source
- Simple arrangement with compact structure
- There is no current flow or Thermal shock
- Food is boiled slowly so degradation of food does not occur.

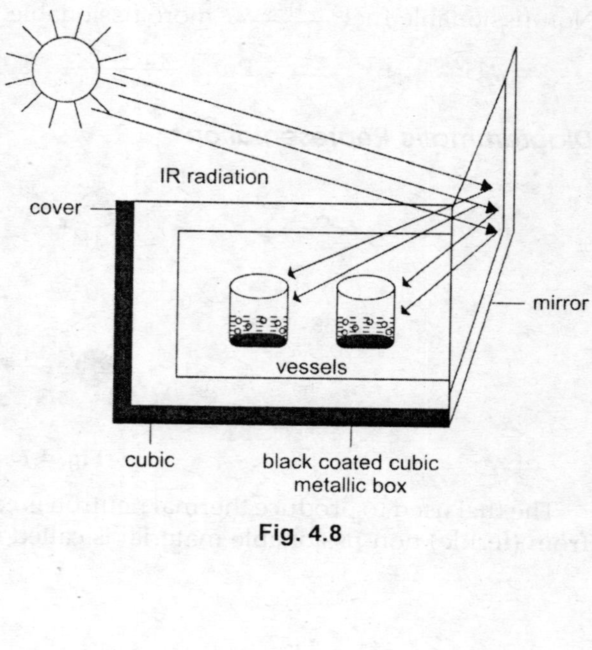

Fig. 4.8

Drawback

- It must be kept towards sunlight.

Solar Water Heater

Instrument	–	Solar water heater
Energy source	–	IR radiation of sun
Conversion	–	Thermal conversion
Principle Used	–	Black body radiation and reflection

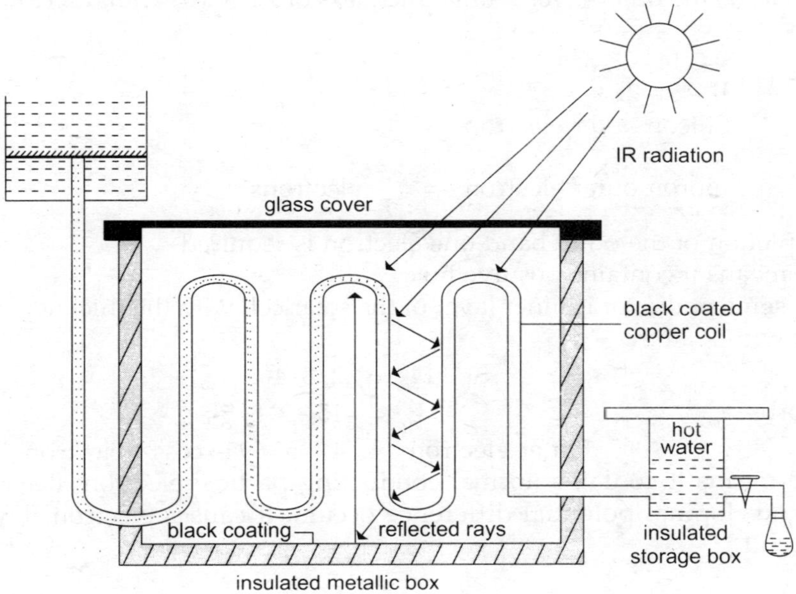

Fig. 4.9

Arrangement

- Bottom was coated black
- This box is covered by a glass plate
- Sunlight is passed inside
- Inside circulated black coated copper coil is kept
- Because of radiation reflection, water is heated
- This hot water is saved inside

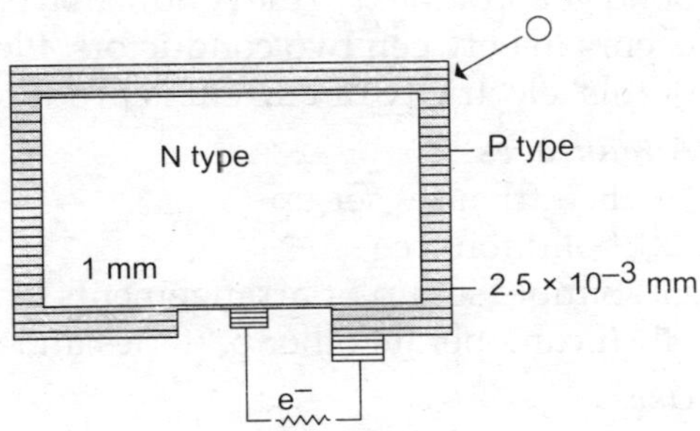

Fig. 4.10

4.15 PHOTOVOLTAIC CELL OR PHOTO GALVANIC CELL OR PHOTO CONVERSION

Instrument	–	Solar cell or Photo Galvanic cell or Photo voltaic cell
Energy Source	–	Solar energy
Conversion	–	Photo conversion
Principle	–	Photovoltaic effect

Solar rays attacked on P-semiconductor and N-semiconductor. Because of this, potential differences occur. Due to this potential difference, the electron flows from N-type to P-type. This is called photovoltaic effect.

Photo light causes voltage difference between two layers, this is called photovoltaic effect.

Solar Cell Construction

Solar cell consists of two different thickness of semiconductor layers. P-type semiconductor is the outer layer with a thickness of 2.5×10^{-3} mm. It is Silicon doped with Boron.

Silicon Si – 14 – 2, 8, 4,

Boron B – 5 – 2, 3

Silicon outer electron = 4

Boron outer electron = $\frac{3}{7}$ electrons

For completion of the outer band one electron is required

P – type means it contains positive hole.

N – type semiconductor is inner layer of the solar cell with the thickness of 1 mm. it is Si doped with P atom.

Silicon Si – 14 = 2, 8, 4

Phosphorus P = 15 - 2, 8, 5

Outer electron = 4 + 5 = 9 excess 1 electron

N - type means it contains normal conducting particle electron. Because of this thickness and doping, potential difference occurs. It causes electron flow through external circuit.

Function

Solar rays fall on the top layer P - type semiconductor. The electrons from the valence bond get promoted to the conduction band. It crosses P - N junction. Potential difference occurs in between two conductors. Electrons flow from one layer to another. Because of this, electricity or current is produced.

Advantages

1. Eco-friendly set-up
2. Pollution free
3. Simple set up or arrangements
4. It can operate other batteries and storage devices

Uses

1. Street lights are operated by solar cells.
2. Water pump, robs, calculators, radios, TV are operated by solar cells.
3. Vehicles are operated by solar cells.
4. Space craft, satellites are operated by solar cells.

Applications

- Solar battery (Street Light)
- Solar cells are used for streetlights operation.

- Simple setup for electricity production.
- While solar light radiation are falling on the surface, streetlights are not operated. During the night time, street lights are automatically switched on

By the same method a lot of solar cells operated.

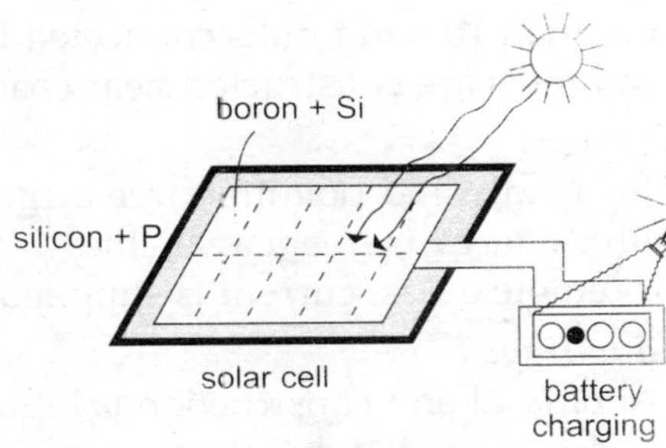

Fig. 4.11

Solar battery

- If more solar cells are connected in series, it is called solar battery
- Solar batteries are used to produce higher electricity
- Solar batteries are used in hill stations and remote villages.

Uses of solar battery

- Pollution free and eco-friendly set-up used in forest and hill stations
- Using solar batteries, vehicles and radio are operated
- Using solar batteries, satellites and spacecrafts are operated
- Some electronic devices, for an example calculator watches are functioning
- Solar energy can be stored in rechargeable batteries

Fig. 4.12: Solar battery

4.16 WIND ENERGY

Air moving very fast from higher to lower attitudes is called wind. The energy created by fast moving air is called wind energy. The wind energy is converted to electric energy by Windmills. The wind energy is created by solar energy (sun).

Energy Source	:	Wind energy
Energy conversion	:	Wind energy is converted to electrical energy
Converter	:	Windmills

Construction and operation

A very big tower is constructed from the ground. Top of the tower is connected with three propeller blades. Propeller blades are made by polymer composites. Wind rotates the blades. During rotation, as wind energy increases, automatically gear increases. This wind energy is converted to electrical energy.

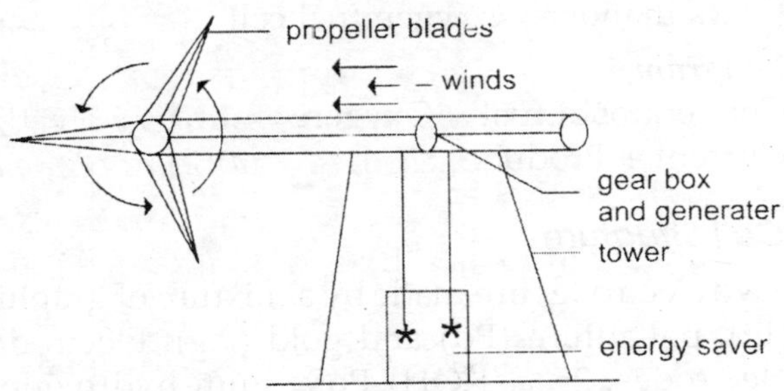

Fig. 4.13

Energy conversion

Sun energy ⟶ Wind energy ⟶ Mechanical energy ⟶ Electrical energy

- Energy generated by a wind mill is of 100 kw power
- Speed of wind mill is 15 km/hr

- More than 10 wind mills connected in the correct series form "wind farms"
- Wind mills are constructed near coastal areas, hills and open grasslands

Advantages

- Eco-friendly and pollution free arrangement
- Source energy is freely available
- For remote areas, current is supplied by windmills

Disadvantages

- Cost of land and construction is high
- We cannot predict the direction of wind
- It affects wireless and television signals while rotation.

Power or Electricity Saving Devices

Electrolytic and electrochemical cells are power saving devices. The cells are connected in series to form batteries. In cells, chemical energy is converted to electrical energy. Based on the chemicals, cells are classified into three types:

Fuel cell (Gas/liquid)
Electrochemical cell (Solid/liquid)
Electrolytic cell (Solid/solid)

4.17 FUEL CELL

Principle

Fuel cell is a device, which converts chemical energy into electrical energy by conventional fuel. Conventional fuels are H_2 (hydrogen), propane and methyl alcohol. Oxygen is the main energy-generating source.

Example for fuel cells:

(a) Hydrogen – Oxygen fuel cell
(b) Propane – Oxygen fuel cell
(c) Methanol – Oxygen fuel cell

Diagram

Conventional fuel + Oxygen ⟶ Current + Product

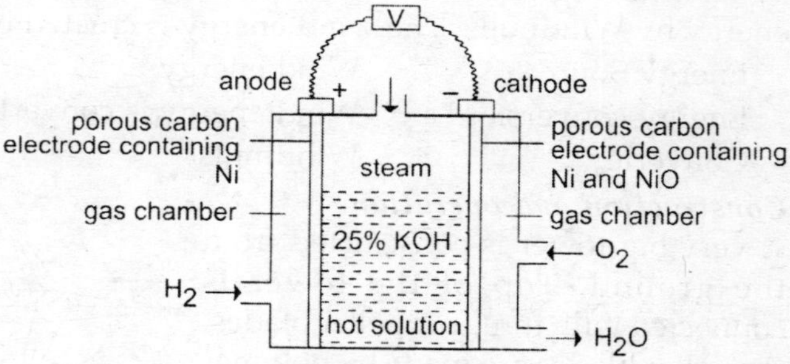

Fig. 4.14

Cell Structure

Two electrodes are made by a mixture of graphite with nickel powder (Ni) or platinum (Pt), palladium (Pd) and gold (Ag). Electrodes conduct electrons. In between two electrodes, 25% (KOH) Potassium hydroxide solution is kept inside. It acts as an electrolyte. Outside the cell, two gas inlets are present. One steam inlet and product outlet are present at the top and bottom of the cell. Oxygen gas passe through the cathode compartment and Hydrogen gas passe through anode compartment at the same time. From this cell, 100% pure water comes outside. 70–80% of chemical energy is converted to electrical energy.

Reaction

Anode (oxidation)	$H_{2(g)} + 2OH^-$	$\longrightarrow$	$2H_2O + 2e^-$ $E^\circ = -0.828$
Cathode (reduction)	$1/2O_2 + H_2O_{(l)}$	$\longrightarrow$	$2OH^-$ $E^\circ = 0.401V$
Overall reaction	$H_2 + 1/2O_2$	$\longrightarrow$	H_2O

$$\begin{aligned}\text{Total Current} &= E^\circ_{cathode} - E^\circ_{anode}\\ &= 0.401 - (-0.828)\\ &= 01.401 + 0.828 = 1.229V\end{aligned}$$

Advantages of Fuel Cells

1. A fuel cell has high efficiency
2. Maintenance of the fuel cell is easy.
3. The chemical process involved in it is clean. It does not produce polluting exhaust
4. The fuel consumption in it goes to completion and it is fuel-efficient. 70% of chemical energy is converted
5. Electrodes need not to change often
6. No harmful products are formed in fuel cell.

Applications and Uses

1. Water is a product from fuel cell. It is in 100% pure state. This water is used in space crafts
2. Electricity produced and used in spacecrafts, submarines and army vehicles

Drawbacks

1. Construction is tough and gases cannot be stored everywhere
2. Commercially fuel cells are not used in households.

4.18 BATTERIES

A battery is a device that converts chemical energy into electrical energy. A battery is a device in which the cells are arranged in series. It gives direct electric current at a constant voltage.

Types of Battery

Based on the shelf life and reproducibility, batteries are classified into two types:

Electrode: Electricity conducting metal rod

Electrolyte: Electricity conducting solution.

Flow chart 4.4

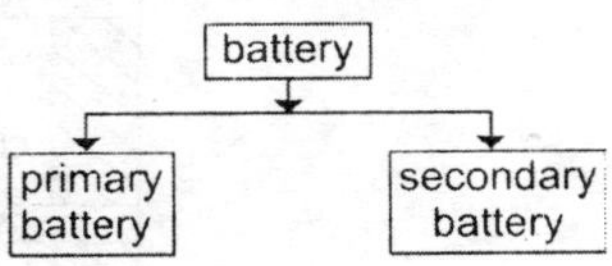

Primary Battery

The primary cell is a device that converts chemical energy in to electrical energy, but chemicals are unable to reproduce by external current and are not rechargeable. Simply explained, reactions are not reversible.

The cell or battery, which are not rechargeable, is called primary battery.

Example, Dry Cells and Mercury Cells

Uses

Primary cells are used in radio, calculator and torch light. The total set-up is covered by a thick paper (carton).

Secondary Battery or Secondary Cells

The Secondary cell is a device that converts chemical energy in to electrical energy, but chemicals are reproduced by external current and are rechargeable. Simply explained, reactions are reversible.

The rechargeable storage devices are called secondary battery.

Example:

Lead - Acid battery used in vehicles

Nickel - Cadmium battery used in calculators and Electronic camera flashes

4.19 ALKALINE BATTERIES

Type of battery	:	Primary battery.
Nature	:	Not rechargeable.
Chemicals	:	Zinc powder, MNO_2 and KOH
Electrode	:	Graphite rod
Alternate Name	:	Dry cell, heavy-duty battery

Construction

- A battery contains zinc cylinder (Anode) and at the centre a graphite rod is fixed (Cathode).
- The cylinder is filled with zinc powder, manganese dioxide (MnO_2) and potassium hydroxide (KOH).
- It conducts electricity.
- The total set-up is covered by thick paper.
- Upper side and lower side are covered by a metal cap.
- EMF of this cell is 1. 5 V.

Diagram

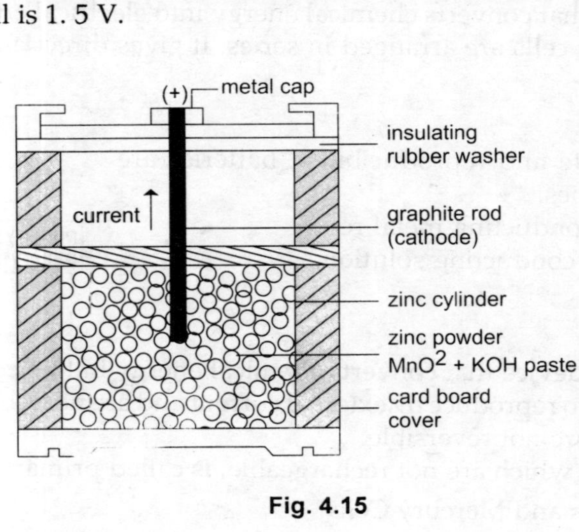

Fig. 4.15

Cell Reaction

Anode (oxidation)	$Zn_{(s)} + 2OH^-$	$\longrightarrow$	$Zn(OH)_2 + 2e^-$
Cathode (reduction)	$2MnO_{2(s)} + 2e^- + H_2O$	$\longrightarrow$	$2OH^-_{(aqu)} + Mn_2O_{2(s)}$
Overall reaction	$2MnO_{2(s)} + Zn_{(s)} + H_2O$	$\longrightarrow$	$Zn(OH)_2 + MnO_{2(s)}$

Advantages of Alkaline battery

1. Zinc metal does not react with based (OH^-) chemicals
2. Shelf life of alkaline battery is high
3. Zinc powder is used as Anode. So it provides more electrode area
4. It performs better in cold weather

Uses: It is used in torch light, wall clock, radio.

Difference between Dry cell (Lechlanche cell) and Alkaline battery

No. Dry Cell (Lechlanche cell)	Alkaline Battery
1. In this cell MnO_2, NH_4Cl and $ZnCl_2$ mixture is used	In this cell Zinc powder, MnO_2 and KOH mixture is used
2. Shelf life less	Shelf life more
3. It cannot used for a heavy battery	This is a heavy-duty battery
4. Without use, efficiency decreases	That is not the case here

4.20 LEAD ACID BATTERY OR LEAD STORAGE CELL

Type of battery : Secondary Cell
Nature : Rechargeable
Chemicals : Lead dioxide (PbO_2) and Sulphuric acid (H_2SO_4)
Electrode : PbO_2 plate (Cathode)
Pb plate (Anode)
Cell notation : $Pb/PbSO_4 // H_2SO_4 (aq) / PbO_2 / Pb$

Construction

- It is made up of a cubic polymer box
- Inside PbO_2 (lead dioxide) plates are kept and connected in series. It is the cathode.
- On the other side Pb (lead) plates are kept and connected in series. It is the anode.

Diagram

The dilute sulphuric acid is filled up to the mark. Two different plates are kept in an electrolyte separately. These are separated by insulator. Sulphuric acid dilution is 20 to 21% and density is 1.3 g/ml. Inside 3 – 6 sets are connected and electricity produced by the cell is 12V.

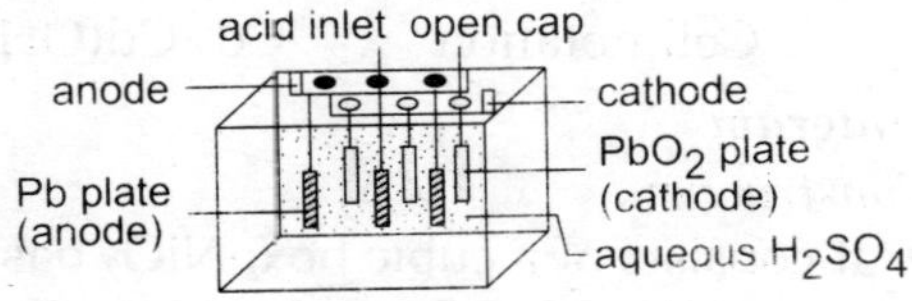

Fig. 4.16

Operation

- Chemical energy is converted to electrical energy
- Based on the cells, electricity is produced
- During the usages, it acts as a voltaic cell or an electrochemical cell
- During recharging. It acts as an electrolytic cell

Reactions During usage

Anode reaction

$$Pb_{(s)} \longrightarrow Pb^{2+}{}_{(aq)} + 2e^-$$
$$Pb^{2+} + SO_4{}^{2-} \longrightarrow PbSO_{4(s)}$$
$$Pb_{(s)} + SO_4{}^{2-}{}_{(aq)} \longrightarrow PbSO_{4(s)} + 2e^- \quad \ldots 1$$

Cathode reaction

$$PbO_2 + 2e^- + 4H^+ \longrightarrow Pb^{2+} + 2H_2O$$
$$Pb^{2+} + SO_4{}^{2-}{}_{(aq)} \longrightarrow PbSO_{4(s)}$$
$$PbO_2 + 4H^+ + SO_4{}^{2-} + 2e^- \longrightarrow PbSO_{4(s)} + 2e^- \quad \ldots 2$$

Overall cell reaction (discharging)

$(1) + (2) \Rightarrow$

$Pb_{(s)} + PbO_{2(s)} + 4H^+ + 2SO_4{}^{2-} \rightarrow PbSO_4 + 2H_2O + \text{Energy}$

Recharging Reaction

Before recharging, sulphuric acid is refilled and density is checked. Density is measured by hydrometer.

Reaction

$PbSO_4 + 2H_2O + \text{External Current} \longrightarrow Pb_{(s)} + PbO_{2(s)} + 4H^+ + 2SO_4{}^{2-}$

Uses

1. It is used in cars, buses and other vehicles
2. It is used in SI engines for fuel ignition purpose
3. It is used in invertors

4.21 NICKEL - CADMIUM CELL OR NICKEL CADMIUM BATTERY

Type of battery	:	Secondary battery
Nature	:	Rechargeable
Chemicals	:	Cadmium, alkali solution (KOH) and Nickel oxide
Electrodes	:	Nickel oxide (Cathode) and Cadmium (Anode)
Cell notation	:	$Cd/Cd(OH)_2$ // KOH (aq) / NiO_2/ Ni

Diagram

Construction:

- It consists of a cubic box, NiO_2 paste is filled in the Ni plate box. It is connected in series. It acts as the cathode
- Cadmium electrodes connected in series. It acts as the Anode
- KOH solution is filled in the cell. It acts as an electrolyte
- This cell produces 1.4 V electricity
- From this reaction $Cd(OH)_2$ precipitates

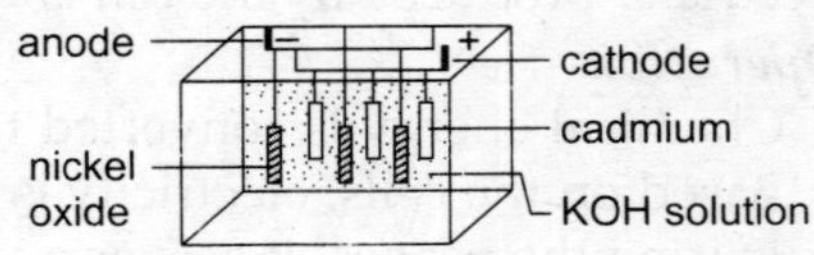

Fig. 4.17

Discharging Reaction

Anode	$Cd_{(s)} + 2OH^-$	$\longrightarrow$	$Cd(OH)_2 + 2e^-$
Cathode	$NiO_2 + 2H_2O + 2e^-$	$\longrightarrow$	$Ni(OH)_2 + 2OH^-$
	$Cd + NiO_2 + 2H_2O$	$\longrightarrow$	$Cd(OH)_2 + Ni(OH)_2$ + Enerty

Both products are insoluble.

Recharging Reaction

Current passes in the opposite direction. The reversible reaction takes place.

$$Cd(OH)_2 + Ni(OH)_2 + \text{Current} \longrightarrow Cd + NiO_2 + 2H_2O$$

Advantages

- It is small and compact
- It can be constructed in a sealed container
- Shelf life of the battery is high.

Uses

It is used in calculators, electronic camera flashes and rechargeable flash lights, sensors, etc.

Drawback

- Nickel and cadmium are hazardous chemicals
- During disposal of the cells these chemicals enter the earth and enter the human body. It causes cardiovascular diseases.

4.22 LITHIUM BATTERY

Reason for Lithium Battery

Other batteries are constructed using a paste of metal oxides. Because of temperature changes, batteries are affected; hence lithium batteries are developed. In this cell all components are in the solid state.

Advantages of lithium battery (it is called future battery):

- In this battery, electrolyte is in the solid state.
- Lithium is a lightweight metal.
- To produce one mole of electrons, 7 g of lithium is required.
- Lithium has the higher anodic potential.
  emf = –3.05 V. Therefore, it generates higher voltage current.
  E cell = ER – E left = ER + 3.05 V
- In this battery, anode, cathode and electrolyte are in the solid state. So there is no risk and it is not affected by weather.
  Example for Lithium battery
  Li /TiS_2 and Li/S

Lithium Titanium sulphide battery or Lithium Solid battery

Type of battery	:	Secondary battery
Nature	:	Rechargeable
Chemicals	:	Lithium and Titanium sulphide
Electrodes	:	Lithium – Anode

Titanium sulphide : Cathode
Electrolyte : Polymer solid electrolyte
Voltage : 3V

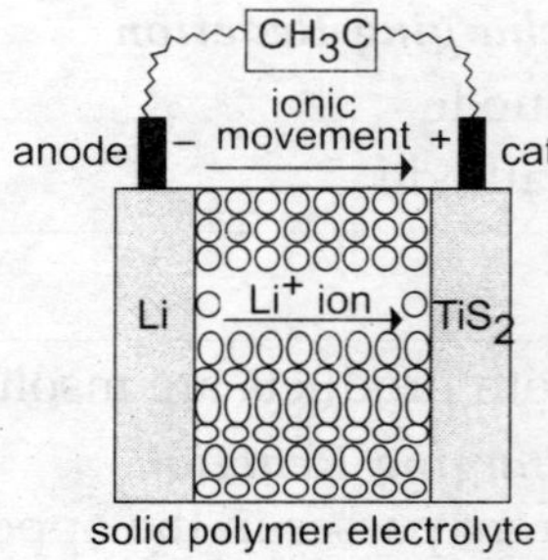

Fig. 4.18

Construction

- Chemicals are packed inside small sized polymer containers or metal containers.
- Electrolyte allows only the ions to move from anode to cathode
- Moving electrons form the external circuit.

Reactions

Anode	$Li^+_{(s)} \longrightarrow Li^+ + e^-$	Eo = –3.05 V
Cathode	$TiS_{2(s)} + e^- \longrightarrow TiS_2^-$	
Net reaction	$Li_{(s)} + TiS_{2(s)} \longrightarrow Li^+ + TiS_2^-$	

During the formation of TiS_2^-, Li^+ ion move from anode to cathode through polymer electrolyte

$$Li^+ + TiS_2^- \longrightarrow LiTiS2$$

Rechargeable reactions

$$LiTiS_2 \rightleftharpoons Li^+ + TiS_2^-$$
$$TiS_2^- \longrightarrow TiS_2 + e^-$$
$$Li^+ + e- \longrightarrow Li$$

Chemicals are reproduced.

Uses

It is used in cell phones, laptops and other electronic devices

Lithium sulphur battery

Type of Cell : Secondary battery
Nature : Rechargeable
Chemicals : Lithium, Sulphur, β-Alumina
Electrode : Two graphite rods
Electrolyte : β-Alumina (solid)

Construction

- A thin walled container, filled by sulphur at molten state and with one graphite rod partially immersed in it acts as the cathode
- In the same container β-Alumina container is kept like the "U" shape
- The "U" shape container is filled with lithium at molten state
- β–Alumina separates lithium and sulphur
- β–Alumina acts as an electrolyte.

Reactions

Anode	$2Li_{(s)} \longrightarrow 2Li^+ + 2e^-$
Cathode	$S + 2e^- \longrightarrow S^{2-}$
Reaction	$2Li + S \longrightarrow Li_2S$

In cathode, S_2 forms. This is not stable. Sulphur molecular formula is S_8. So, it tries to form the stable molecule.

$$S^{2-} + ns \longrightarrow \left[S_{(n+1)}\right]^{2-}$$

Polysulphite ion

Lithium ions move from "U" shape container to molten sulphur container through β-Alumina electrolyte.

Uses

- It is used in electric vehicles
- Rechargeable emergency light, etc.
- It has a high energy density

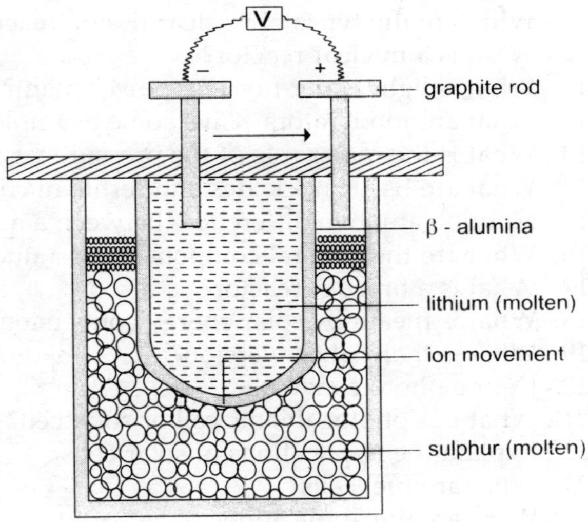

Fig. 4.19

GLOSSARY

Fossil Fuel: Naturally biodegraded plant and animal products are called Fossil Fuels.
Radioactivity: Emission of powerful radiation continuously by the deformation of an element.
Chemical reaction: Valence electrons are changed.
Nuclear reaction: Nucleus is changed
Thermal neutrons: Slow neutrons
Critical neutrons: Slow neutrons.
Critical Mass: The minimum amount of fissionable material required for spontaneous chain reaction.
Nuclear fission energy: Energy emitted from nuclear fission reaction.
Nuclear fusion energy: Energy emitted from nuclear fusion reaction.
Nuclear pile: Where the lot of material is inserted.

Assessment Questions

Answer all questions (2 marks)

1. Define nuclear fission.
2. Mention a few important characteristics of nuclear fission.
3. What is nuclear fusion? Give an example.
4. Give any two differences between nuclear fission and fusion.
5. What is nuclear chain reaction?
6. What is super critical mass and sub-critical mass of U^{235}?
7. What is nuclear energy?
8. Give any one nuclear fission reaction, mention the factors that impede the nuclear chain reaction.

9. What are the types of nuclear fission reaction?
10. What is a nuclear reactor?
11. What is light water nuclear–power plant?
12. What are moderators? Give some examples.
13. What is the major role of the pressure vessel in the nuclear reactor?
14. What are fissile nucleides and fertile nuclides?
15. Mention any two differences between a nuclear reaction and a chemical reaction.
16. What are the general components of a nuclear reactor?
17. What is a breeder reactor?
18. What is meant by solar energy conversion? How is it done?
19. What is thermal conversion?
20. Define photo conversion?
21. What is a photogalvanic cell or solar cell?
22. Explain the applications of solar cells.
23. What are fuel cells?
24. What are the applications of $H_2 - O_2$ fuel cell?
25. What is wind energy? How is it obtained?
26. What are the drawbacks of wind energy?
27. What is a battery? How does it differ from a cell?
28. What are the important requirements of a battery?
29. What is a primary battery? Give an example. Or What are primary cells?
30. What are secondary cells?
31. Write the overall equation for the reaction taking place in an alkaline battery.
32. What are the advantages of alkaline battery over dry battery?
33. Write the cell representation of lead storage cell?
34. Write the charging and discharging reactions of a lead accumulator.
35. What are the applications of lead acid storage cell?
36. Write the cell representation of NICAD battery?
37. What are the advantage and disadvantages of NICAD battery?
38. How are anodic and cathodic electroactive materials made in Ni-Cd battery? Or How is NICAD battery constructed?
39. Describe lithium battery?
40. What are the advantages of Li-S battery?
41. Lithium battery is the cell of future, why?

Answer all questions (16 Marks)

1. What is a nuclear reactor? Describe the components of a light-water nuclear power plant with a suitable block diagram.
2. With a neat sketch explain the functioning of a nuclear reactor.
3. Write in brief about chain reactions and nuclear fission reactions.
4. What are breeder reactors? Explain with an example.
5. What are nuclear chain reactions? Explain how to improve the amount of nuclear energy with illustration.
6. State the principle and applications of solar batteries.
7. Describe the construction and working of $H_2 - O_2$ fuel cell.

8. What are fuel cells? Explain the construction and working of fuel cells.
9. What is primary battery? Give an example.
10. Write a note on Ni-Cd battery.
11. Write a brief note on lead-acid storage cell.
12. Explain the construction and workings of a lead acid battery.
13. Write notes on lithium battery.
14. What is reversible battery?
15. Describe the construction and working of a lead acid storage battery with reaction occurring during charging and discharging.
16. How is NICAD battery constructed? Explain with cell reaction.
17. Give the description of lead storage cell and explain its functioning during discharging and recharging.
18. Give an account of Lead-acid and Ni-Cd batteries.
19. Explain the construction and working of a lead storage battery.
20. i. What is a breeder reactor? Explain its principles.
    ii. Write notes on wind energy.
21. i. Explain various methods involved in solar energy conversion.
    ii. What are the difference between nuclear fission and nuclear fusion?
22. i. Write a note on fuel cells.
    ii. Explain the construction and working of lead acid battery.

5

Engineering Materials

5.1 INTRODUCTION

The materials which are made up of chemicals and used for engineering purposes are called engineering materials. They may be metallic or non-metallic components. The raw solid substances used to make products and structures, and the science related to the study and treatment of these substances and their applications. All fields of engineering have chemistry, physics and mathematics as their basis.

Flow chart 5.1

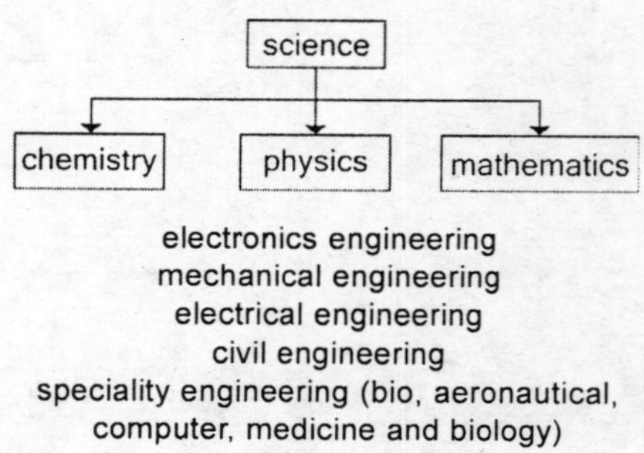

Chemistry

The science which deals composition of matter and reactions is called chemistry.

Some important industrial engineering materials

1. Abrasives
2. Refractories
3. Lubricants
4. Carbon Nano tubes

5.1.1 Abrasives

An abrasive is a substance that can be used to grind another substance. It can also be used for cutting, surfacing, finishing and shaping another material. It can be classified by a high melting point, high hardness and chemical inertness.

5.1.2 Important Properties

Abrasives are strong and hard materials. It should not be affected by the frictional heat. It should possess high refractoriness and resistance to mechanical shock. It should be tough and sharp.

Abrasive Power

The capacity of a particular abrasive to rub down the other objects. It is called its abrasive power.

Hardness

Hardness is the capacity of an abrasive to scratch other objects.

Scale of hardness is Moh's scale

Moh's scale was proposed by a mineralogist

Abrasives are arranged in an increasing order.

Moh's Number	*Name of the Chemical*	*Molecular Formula*
1.	Talc	$3MgO, 4SiO_2, H_2O$
2.	Gypsum	$CaSO_4, 2H_2O$
3.	Calcite	$CaCO_3$
4.	Fluorite	CaF_2
5.	Apatite	$CaF_2, 3Ca_3(PO_4)_2$
6.	Feldspar	$K_2O, Al_2O_3, 6H_2O$
7.	Quartz	SiO_2
8.	Topaz	AlF_3, SiO_2
9.	Corundum	Al_2O_3
10.	Diamond	$C(sp^3)$

Toughness

The abrasives are brittle and sharper edged. It measures the resistance of the material for sharpening (opposite to brittleness).

Refractoriness

The resistance of an abrasive breaking capacity and its grain with the rise in temperature is called refractoriness. That means withstanding capacity of the material at high temperature is called refractoriness.

5.1.3 Classification of Abrasives

Based on hardness it is classified into two types

1. Hard Abrasives
2. Soft Abrasives

Based on availability it is classified into two types

1. Natural Abrasives
2. Synthetic Abrasives

Soft Abrasives: Hardness value is in between 1 – 4 Moh's scale

Example: Red rouge – Fe_2O_3, nH_2O, Green rouge – Cr_2O_3.

Hard Abrasives: Hardness value of above four is called hard abrasives.

Example: Diamond, Garnet and Emery ($Al_2O_3 + Fe_2O_3$)

Natural Abrasives: It is available naturally and has a high hardness value in Mohs scale.

Example: Diamond and Quartz

Synthetic Abrasives: They are artificially prepared hard materials.

Example: Fused Alumina and Boron Carbide

Based on availability, abrasives are again classified into two types. They are also classified by silicon content.

Flow chart 5.2

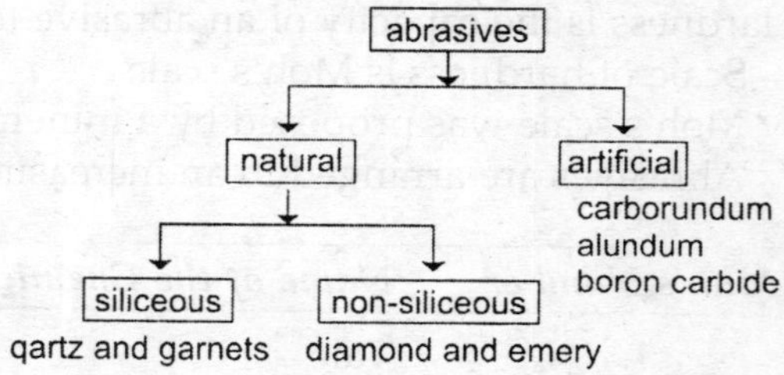

Natural Non-siliceous Abrasives

5.1.4 Diamond

It is pure crystalline carbon. It is a very hard material and its hardness value is 10 Moh's. In diamond, carbon molecules are arranged by sp^3 hybridization. It is chemically an inert and non-conducting material. Based on color, it is classified into two types:

Off Colour – Borts
Black Color – Carbonados

Uses: For drilling rocks, cutting and grinding purpose.

5.1.5 Emery

It is a fine grained, opaque, black colored mineral. Constituents are 55 – 75 % crystalline alumina and 25 – 45 % magnetite. Hardness on Moh's scale is 8.
Uses: For cutting, drilling tools and making abrasive paper.

5.1.6 Corundum

It is a crystalline powder. Constituent compound is Aluminum oxide. Moh's scale value is 9
Uses: For making abrasive wheels in industries.

Natural Siliceous Abrasives

5.1.7 Quartz

It is crystalline silica (SiO_2). It is brittle and has a shell like structure. Moh's scale hardness value of quartz is around seven.
Uses: For making sandpapers and its used in ball mills for grinding pigments (Paint Industry).

5.1.8 Pumacite

It is volcanic dust. It contains SiO_2, Al_2O_3, Fe_2O_3, FeO, H_2O, etc. in it.
Uses: It is used for scratching the surface in many industries.

5.1.9 Garnet

It is a mixture of tri-silicates of alumina, magnesia and ferrous oxide. Moh's value of hardness is 7.5.
Uses: For making abrasive paper and cloth.

Examples of Garnets:

a. $Ca_3Al_2(SiO_4)_3$
b. $Ca_3Fe_2(SiO_4)_3$
c. $Fe_3(SiO_4)_3$

Synthetic abrasives preparation properties and uses

Examples of synthetic abrasives:

1. Silicon carbide or Carborundum (SiC).
2. Boron carbide or Norbide (B_4C).
3. Alundum (Al_2O_3).
4. Calcium carbide (CaC_2).
   Artificial abrasives have uniform chemical compositions and hardness

5.1.10 Boron Carbide or Norbide

Molecular formula: B_4C
Type of Abrasive: Synthetic Abrasive
Reactant chemicals: Boron oxide (B_2O_3) and Coke (Carbon)

Preparation

The mixture of Boron trioxide and coke is heated in an electric furnace to 2700°C.

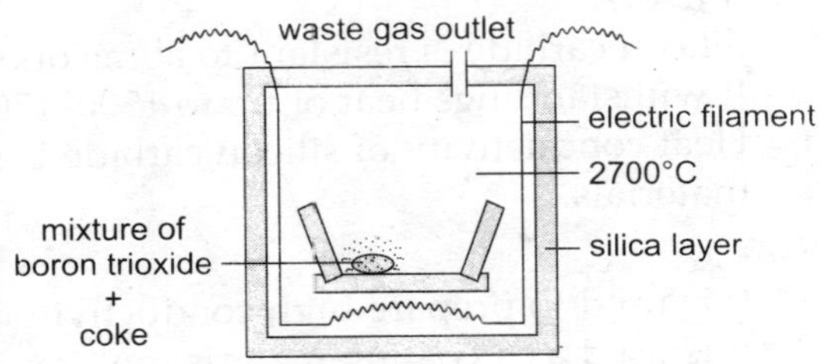

Fig. 5.1

Reaction

$$2B_2O_3 + 7C \xrightarrow{2700^\circ C} B_4C + 6CO$$

Properties

It is a lightweight black colored material and its hardness on Moh's scale value is nine. It can resist oxidation much better than diamond. It is highly inert to chemicals and the hardest material ever synthesized

Uses

- It is used in industries for cutting hard materials
- It is used in crushers.

5.1.11 Silicon Carbide or Carborundum

Molecular formula: SiC
Type of Abrasive: Synthetic Abrasive
Reactant chemicals: Sand (SiO_2) and Coke (C)

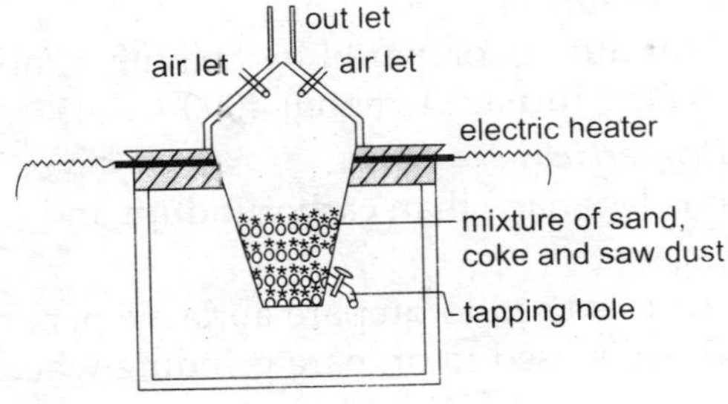

Fig. 5.2

Preparation

In an electric furnace, 60% of sand and 40% of coke are mixed. Then it is charged into an electric furnace. After that, a small amount saw dust, salt are inside the furnace. It is heated in a heater to 1500°C. Finally, silicon carbide is removed from the furnace and mixed with the bonding agent. Then it is shaped, dried and fired. Based on the bonded material, silicon carbide is in the following order:

Self bonded > Silicon nitride bonded > Clay bonded

Reaction

$$SiO_2 + 3C \xrightarrow{1500°C} SiC + 2CO$$

Reason for sawdust and Salt

1. During burning process, sawdust produces gases. This increases porosity on the product.
2. Salt reacts with small amount of iron present in sand. It also produces volatile chlorides. This also increases porosity on abrasive material.

Bonding agents: Clay and silicon nitride.

Properties

1. Silicon carbide has a higher tensile strength, stiffness and hardness.
2. Compared with other ceramics, silicon carbide has the highest thermal conductivity.
3. Because of dimensional stability and polish ability, it is used in precision optical mirrors.
4. Silicon carbide is resistant to abrasion.
5. It withstandings heat of up to 1500–1700°C.
6. Heat conductivity of silicon carbide is intermediate between metals and ceramic materials.

Uses

1. It is used to prepare high conductivity crucibles.
2. It is used as a coating material on graphite tooling.
3. It is used as heating elements in furnaces in the form of rods.
4. It is used as inner walls in chamber kilns, coke ovens, and muffle furnaces.

5.1.12 Alundum

Molecular formula: Al_2O_3

Type of Abrasive: Synthetic abrasive

Required chemicals: Calcined bauxite, coke and iron

Preparation

Alundum is prepared by heating a mixture of calcined bauxite, coke and iron in an electric furnace to about 4000°C.

Properties

It is less hard than carborundum and less brittle and tough.

Uses

1. It is used to prepare abrasive paper
2. It is used to prepare grinding wheel

5.1.13 Applications of Abrasives

Abrasives are used as different states:

1. Abrasive Paper or Cloth
2. Grinding wheels
3. Loose powder

Abrasive Paper or Cloth

Abrasive papers are used to clean the metal surface and make the surface smooth. Before the painting process, the metal surface is polished.

Preparation of Abrasive Paper

Paper is rolled in a roller. This paper or cloth is moved by rolling wheels. From the glue chamber, glue is sprayed on the paper. It is passed and loose abrasive powder to sprayed on the glue sprayed paper. Then, the wet abrasive paper is dried by hot air. After drying the abrasive paper passes through the cutter.

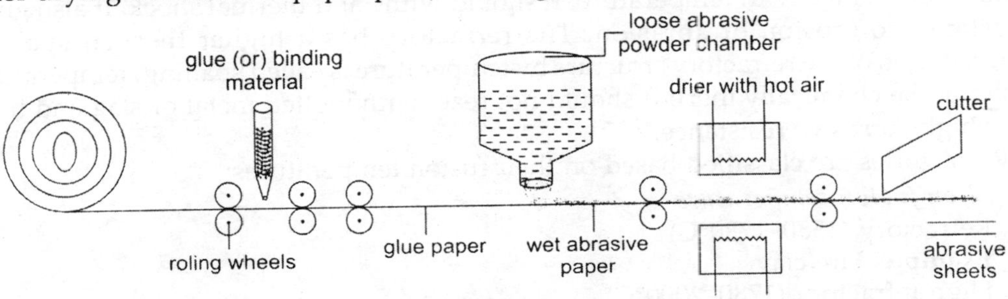

Fig. 5.3

Uses: Abrasives are used for smoothening the surface and polishing the metal.

Example: Silicon Carbide

Grinding Wheel

Manufacture process

Grinding wheel is prepared by using the mixture of abrasive grains with binding material.

The mixture is moulded into a particular shape, dried and heated.

Uses

1. It is used to remove rust from iron surfaces
2. Cutting tools and sharpening the surfaces.

Loose Powder

To clean the surface and for gripping purpose loose powder is used. Using loose powder metal surface coated.

Example, Quartz, Garnet, Pumacite and Norbide.

5.2 REFRACTORIES

5.2.1 Introduction

Engineering materials are prepared at high temperatures. Engineering designs are based on high temperatures and high pressures. So, we have to use high temperatures withstanding materials. Refractory, is defined as the chemical material which will withstand high temperatures without melting. Its fusion temperature is very high and its structure does not change at high temperatures. Refractory is a type of ceramics.

These are made from the high melting oxides like SiO_2, Al_2O_3, CaO, and MgO and etc.Refractories resist heat excellently. Generally, the inner sides of furnaces are lined with refractory to withstand high temperatures. Refractoies are made by different shapes like bricks, crucibles, tubes and granules.

5.2.2 Characteristics of a Refractory

The resistance to fusion (melt) on increasing the temperature is called refractoriness. The value of refractoriness should be high to be a good refractory. The bulk structure should not change with temperature. It should withstand thermal shock. It also gives resistance to erosion or abrasion. The refractory has a higher thermal spalling (degradation of the refractory structure by temperature is called spalling) temperature. It should be chemically inert. It should not react with molten metal or slag and have very high corrosive resistance.

Refractories are classified based on their fusion temperatures:

Based on fusion temperature

1. Refractory (1580–1780°C)
   Example, Fire clay
2. High refractory (1780–2000°C)
   Example, Chromites
3. Super refractory (above 2000°C)
   Example, Zircon

   Refractories are selected based on the furnace operational temperatures.

5.2.3 Types of Refractory

Based on chemical composition and nature, refractory is classified into three types:

1. Acidic refractory
2. Basic refractory
3. Neutral refractory

Acidic refractory: It is made up of acidic components. It is not affected by acidic but affected by basic, ***e.g.:*** silica, alumina and Fire clay

Flow chart 5.3

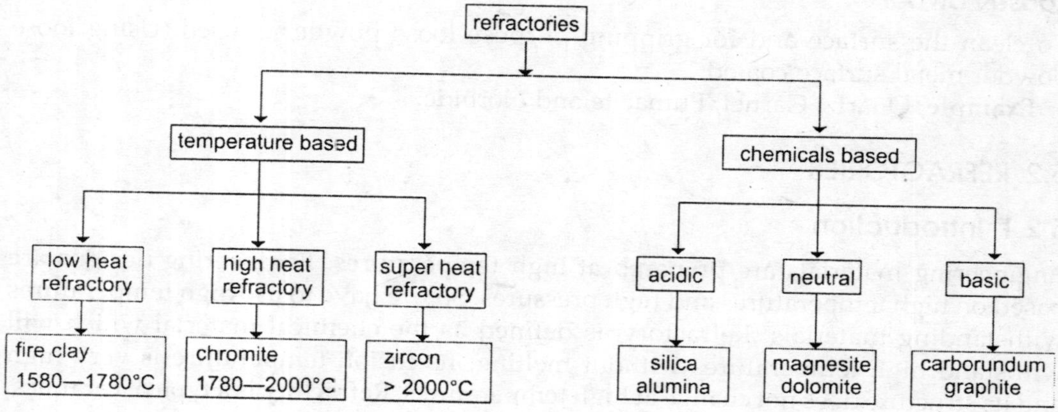

Basic refractory: It is made up of basic components. It is not affect by basic but affected by acidic, ***e.g.*** Magnesite, Dolomite

Neutral refractory: It is made up of weak acidic and basic materials. They are inert towards both acidic and basic substances, ***e.g.:*** Graphite, Chromites, Zirconia and Carborundum

5.2.4 Properties of Refractories

The important properties of refractory with relevance to their utility are

1. Refractoriness
2. Thermal expansion and contraction
3. Spalling resistance
4. Porosity
5. Corrosion and erosion–resistance
6. Load–bearing strength
7. Thermal conductivity
8. Electrical character
9. Chemical characteristics

5.2.5 Refractoriness

Definition: The resistance to fusion on increasing the temperature is called refractoriness. If the temperature increases, material softens and deforms. Refractoriness is measured by the sager cone method. For a refractory, it is the ability of a material to withstand deformation under working conditions.

5.2.6 Applications or Objectives of PCE

- To classify the refractory
- To find the exact softening temperature
- To determine the purity

Method: Seger cone or pyrometric cone equivalent

Principle: The refractories are made from mixtures of several metallic oxides. it is an impure state. Therefore, they do not have sharp melting points. Based on this principle refractoriness is measured. Using refractory material, the standard shape of cone is prepared. The particular temperature at which cone shape is deformed is called softening temperature. If any unknown refractory material equalizes known seger cone softening temperature, then that will be called its PCE number.

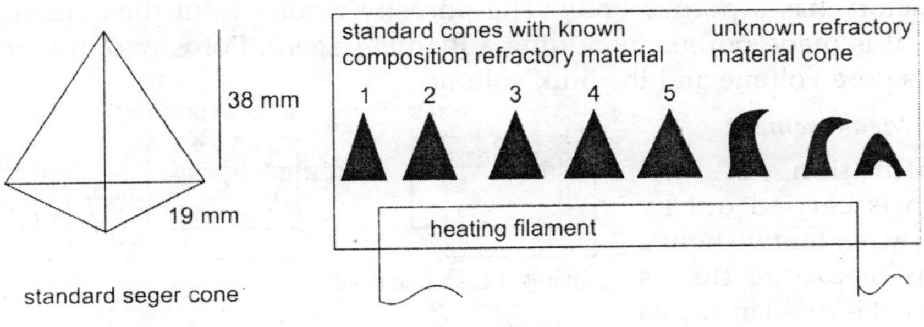

Fig. 5.4

Procedure

Using particular known refractory material, more than one-seager cones are prepared. Then it is numbered using unknown refractory material, seger cone is prepared. Then seger cones are heated gradually by an increasing order of 10°C. At a particular temperature unknown or test cone is deform or soften this is called pyrometric cone equivalent number. Number given for each cone up to 40–45. 1–45 are known refractory material composition and known softening temperature. If test cone softening temperature is equal to standard seger cone number 33, then it is called pyrometric equivalent. Cone number 33 has a softening temperature that ranges from 1700–1750°C.

Standard fire clay has 1700°C. Then test cone may be fire clay.

Cone number	End point or softening temp°C
1	1125
5	1180
15	1410
20	1520

End of the process, the refractory apex bends and touches the surface on which it is kept. Spalling resistance of some refractory is given below.

Silicon carbide > Fire clay bricks > Magnesite > Silica bricks

Thermal expansion and contraction

The expansion and contraction of a refractory with changing temperature is negligible. If it is appreciable, the furnace surface would crack.

Spalling resistance

The breaking, cracking or peeling of a refractory under temperature is called spalling. Sudden increase and decrease in temperature of the furnace is causing thermal spalling. Because of temperature changes uneven contraction and expansion occur. The spalling tendency is directly proportion to the coefficient of expansion.

How to avoid the spalling or remedy

1. Refractory should have high porosity.
2. It should have a lower coefficient of expansion.
3. Good thermal conductivity refractory material should be selected for furnace.

5.2.7 Porosity

The refractory has a porous body. The porosity results from their manufacturing methods. It is made porous by adding a foaming agent. Porosity of the brick is the ratio of its pore volume and the bulk volume.

Porosity Measurement

The saturation of the specimen is carried out by boiling in water for two hours. Weight is measured that is "B"g. Then it is dried in air, its weight is "A"g.

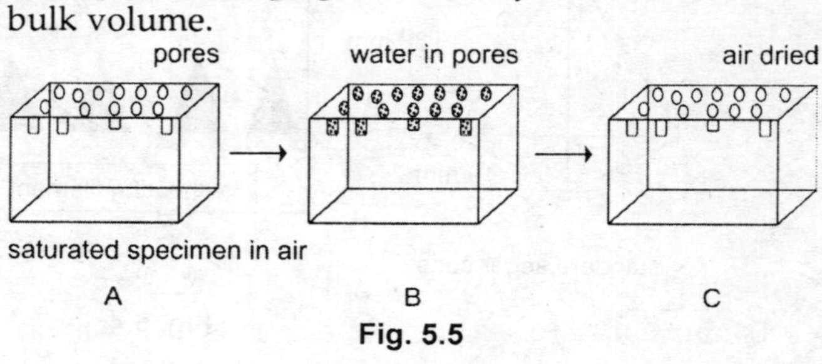

Fig. 5.5

$$\text{Porosity ratio} = \frac{\text{pore volume}}{\text{bulk volume}} \times 100$$

$$\text{Porosity} = \frac{X - A}{X - B} \times 100$$

Advantages of Porosity

The strength of the refractory brick is influenced by porosity. Pores act as insulators and give thermal insulation to the refractory. It decreases thermal spalling.

Disadvantages of Porosity

In a highly porous brick, molten reactants, slags and gases penetrate and damage the brick.

5.2.8 Load-Bearing Strength

The two essential qualities of a refractory are temperature resistance and load-bearing capacity. The good refractory material should withstand high temperature along with load. Generally, softening temperature decreases with increasing load. The load bearing capacity of a refractory can be measured by RUL test. (Refractoriness under Load). Refractoriness under load of a refractory is the ability of refractory to withstand high temperatures under the influence of maximum load without deformation or breaking.

The refractory under – load test (RUL Test)

- This test is performed to measure the safe upper temperature limit up to which the refractory can be used.
- This test is done in a rectangular container specimen of standard size with 5 cm^2 base and 75 cm height
- 1.75 kg/cm^2 load is applied on the refractory
- It is heated at a constant rate of 10°C/min
- During heating, the specimen will soften and therefore, its height will keep decreasing under the load
- This decrease in height is watched with a ruler placed alongside the specimen
- When the height is decreased by 10% of the original height by 7.5 cm, the temperature is noted. Then RUL is expressed in terms of this temperature.

Dimensional Stability

Dimensional stability is the resistance of a refractory to any volume changes when exposed to high temperatures with time.

These are classified into two type:

1. Irreversible
2. Reversible

Irreversible change: Because of heating refractory material changes its structural dimension. It may contract and expand.

Example, Magnesite bricks shrink on heating, silica bricks expand on heating

Shrink: Magnesite (3.05) ⟶ Perclase (3.54)
(Amorphous) (Crystalline)

Expansion: Quartz $\xrightarrow{870°C}$ Tridymite $\xrightarrow{1475°C}$ Cristobalite
(Crystalline) (α-form) (β-form)

5.2.9 Manufacturing Method of Refractories

Raw materials (Dolomite, Magnesite and Bauxite)

↓

Suitable size

↓

Powder material

↓

Fine, uniform sized particles

↓

Volatile impurities removed powder

↓

Purified raw material

↓

Product for molding

↓

Wet refractory

↓

Dry bricks

↓

Hard refractory bricks

Description

1. Crushing – Dolomite, Magnesite, Bauxite raw materials are crushed together.
2. Grinding – Crushed materials are ground well.
3. Screening – By passing through standard metal mesh, fine and uniform sized particles are collected.
4. Calcinations – Uniform fine powder is heated up to 1200–1700°C temperature in rotary kilns. By this process volatile impurities are removed
5. Mineral dressing – Fine powder is introduced in mineral dressing. By this process all mineral impurities are removed and the refractory raw material is purified.
6. Blending – Using a suitable binding material, pure raw material is made into paste.

7. Moldings – This pasty material is moulded in to particular shape by manual or mechanical pressure, like square or rectangle.
8. Drying – Then this solid is dried under air.
9. Firing – This dried material is fired well and converted into a strong hard refractory.

After completion of the process, refractory are fired to improve the following properties:

1. To remove water of hydration
2. Stabilize and strengthen the refractory structure
   Firing temperature varies with different bricks
   For an example,
   Kaolin – 1700°C
   Basic Bricks – 1870°C

5.2.10 Important Refractory Material

1. Alumina
2. Magnesite
3. Zirconia

5.2.11 Alumina Bricks or Fire Clay

Molecular Formula: Al_2O_3
Nature of Refractory: Acidic Refractory
Constituents: 50% of Alumina (Al_2O_3) mixed with clay binder $XSiO_2$

Manufacture

The raw materials calcined bauxite (Al_2O_3), Silica (SiO_2) and Grog (Clay binder) are mixed together and ground well (grinding, crushing, screening). Then the powder is added with water to make paste. Then the paste is made into a relevant shape (Molding). Then it is dried in air (Drying). Molded and dried bricks are fired in a kiln (oven) at 1200–1400°C for 6–8 hours (Firing). Finally refractory alumina bricks are prepared.

[*Grog:* Aluminium silicate (hydrated) or fire clay Al_2O_3. $2SiO_2$. $2H_2O$]

Properties

- It is an acidic refractory. It is not affected by acidic chemicals
- A lower expansion coefficient
- High temperature and high load bearing capacity
- A higher thermal spalling value
- Gases (CO_2, H_2 and CH_4) do not affect.

Uses

- Based on the Alumina percentage it is classified in to two types:
  High duty bricks containing (75% Al_2O_3)
  It is used in cement rotary kilns, metal melting furnaces (Al, Brass). It is highly used in the steel industry
  Medium duty bricks (50–60% Al_2O_3)
  It is used in soaking pits and furnace walls.

5.2.12 Magnesite Bricks

Molecular Formula: MgO

Nature of Refractory: basic refractory.

Constituents: 50% of calcined, caustic Magnesia binding material or iron oxides are mixed together (mixing)

Preparation

Crushing, grinding and screening:

The raw materials, calcined magnesia and caustic magnesia are crushed, ground and screened.

Pasting, molding and drying

Then it is mixed with water and made it to paste. The paste is shaped and dried in air.

Firing

Finally, the air-dried material is fired.

Properties

- Magnesite bricks are basic refractories
- It withstands up to 2000°C without load and 1500°C with 3.5 kg of load
- It is not affected by basic chemicals
- Higher thermal spalling and good strength
- Poor resistance to abrasions.

Uses

- It is used to heat basic materials
- It is used in the steel industry
- It is used in copper converters and reverberate furnaces.

5.2.13 Zirconia Bricks

Molecular formula: ZrO_2

Nature: Neutral

Manufacture:

Crushing, grinding, screening and calcinations

Zirconite mineral (ZrO_2) and semi solid Zirconia or alumina as binders are mixed together and processed. The heated material is mixed with water and made in to a paste. After molding, drying and firing, the wet zirconite mixture is heated up to 1700°C. Zirconite volume changes can be stabilized by MgO.

Properties

- It is a neutral refractory.
- Its thermal expansion temperature is low.
- It is resistant to thermal shock.
- It is affected by acids.

Uses

- It is used in high frequency electric furnaces.

5.3 LUBRICANTS

5.3.1 Introduction

In any type of machinery, there is constant rubbing on surfaces of moving bodies. Because of there develops a resistance to their movement. The resistance to flow smoothly is called "friction". This friction force heat energy forms in between the surface. Because of this heat energy, smooth surfaces are scratched and it causes breakage of the moving bodies. Due to friction, machine efficiency decreases.

5.3.2 Definition for Lubricant

The substance which reduces friction between two moving surfaces is called lubricant. or Lubricant is a substance which reduces the friction force in between two moving surfaces.

Lubrication: By applying lubricant friction reduces between two moving surfaces. This process is called lubrication. or Lubricant applied process is called lubrication or friction-reducing process is called lubrication.

5.3.3 Role of Lubricant

1. It reduces or minimizes friction between two surfaces
2. It prevents direct contact between two surfaces
3. It reduces loss of energy
4. It helps in reducing wear, tear and surface deformation
5. It acts as a seal and corrosion resister
6. Lubricants act as a shock absorber and a coolant
7. It improves the efficiency of machine
8. It reduces the maintenance cost of machine
9. It reduces energy wasted as heat energy is reduced
10. It prevents the expansion of moving metal surfaces.

5.3.4 Types of Lubricants

Based on the physical state, lubricant are classified into following types:

1. Liquid Lubricants, *e.g.* Oils
2. Semi solid Lubricants, *e.g.* Greases
3. Solid Lubricants, *e.g.* graphite
4. Emulsions, *e.g.* Cooling Liquids
5. Gases, *e.g.* N_2

Liquid Lubricants

1.	Animal oils	Whale oil, Lard oil, Tallow oil
2.	Vegetable oils	Olive oil, Palm oil, Castor oil, Rape – seed oil
3.	Mineral oils	Petroleum oils
4.	Synthetic oils	Silicones, Poly glycol ethers
5.	Blended oils	The additives are added with mineral oils

Semi Solid Lubricants: Nowadays semi solid lubricants are used in many fields, because of its stability, *e.g.* Grease and Vaseline.

Solid Lubricants: In an electric instrument, solid lubricants are used.
Example, Graphite, Molybdenum disulphide

Emulsions: Two immiscible liquids are mixed together.
Example, water in oils, cooling liquids, oils in water cutting emulsions.

Gases: In between two surfaces, gases are passed to reduce the friction.
Example, N_2 gas.

Properties of Lube Oils or Lubricating Oils

Animal oil or Vegetable oil

These oils are glycerides of higher fatty acids and are used to prepare grease. It is used as a "blending agent". Lubricating oils are used as additives. They have good oiliness.

Drawbacks

They are costly and oils undergo oxidation easily. They undergo decomposition on heating. Fatty acids are hydrolyzed by moisture.

Petroleum oil or Mineral oil

Mineral oils are C_{12}–C_{50} carbon content oils. Petroleum oil (lube oil) containing hydrocarbons. Nevertheless, it possesses poor oiliness, oiliness can be improved by addition of animal and vegetable oil. The residue is obtained from crude oil by the distillation process. However, it cannot be used as such. Because of the impurities.

Impurities: Wax, Asphalt (Tar), Sulphur, colored substances and other oxidisable impurities.

5.3.5 Mineral Lubricating Oil Refining Method

- Lubricating oil is purified by the following methods:

1. De-waxing

Removal of wax from mineral oil is called de-waxing. Waxes are removed at lower temperatures. Mineral oil is dissolved in a suitable solvent. Then wax is removed by the chilling process and filtration. Solvent is removed by distillation and it is reused.

Solvent: Benzene and Propane

2. Removal of Asphalt

Asphalted and naphthalene materials are removed by acid refining process. De-waxed oil is treated with concentrated. H_2SO_4 (sulphuric acid) and is mixed well. Some impurities dissolve in acid and it becomes sludge. Sludges is filtered and removed. Then is filtered oil is neutralized by NaOH solution. The oil layer is separated from the aqueous layer.

3. Desulphurization

Sulphur removal process is called desulphurization. Oil treated with Ni/H_2 and Sulphur is removed as H_2S and the unsaturated compound is converted to the saturated compound

4. Decolourisation

After the above purification process, oil is treated with fullers earth (poly silicate) or activated clay. Colour is removed from oil and this process is called decolourisation. During depolarization micro particles are also removed.

Effect of Asphalt

If asphalt is present in oil it deposits on the engine parts. It increases friction. That's why it should be removed from oil.

5.3.6 Synthetic Lubricants Oils

Need for synthetic lubricant

1. Under conditions of high machinery and wide variations in temperature conditions, other lubricating oils are not suitable.
2. For the industrial and military needs of lubricants, the conditions will be too demanding for petroleum oils

To rectify this problem synthetic lubricants are prepared.

Properties of Synthetic Lubricants

1. Synthetic lubricant oil perform well, even below 26°C and above 121°C
2. It is easily prepared by synthetic methods
3. Molecular structures that are stable under specific operating conditions
4. These lubricants have low pour point and high flash points
5. They have high viscosity indices

Example:

2-ethyl-hexanol is used in jet engines and military equipments
Polyglycol ethers
Silicate esters
Silicones

5.3.7 Properties of Silicone Oils

1. These are high temperature lubricants
2. It is used in equipment manufacturing industries
3. It has high thermal stability
4. It has fire resistivity.

Blended Oil or Compounded Oil

1. Single oil cannot act as an ideal lubricant for many of the modern machineries
2. So using additives, petroleum oil character is improved

Important Additives and Functions in lubricating oils:

No.	Improved Property	Example for an Additive	Use or Function
1.	Anti-oxidant	Aromatic, Phenolic and Amino compounds. [β-naphthol, diphenyl methane]	It is used in I.C engines It reduces oxidation during friction
2.	Corrosion inhibitor or Corrosion preventor	Organic Antimony or Organic phosphorus compounds [Tri cresyl Phosphate]	It reduces metal degradation of surfaces and corrosion
3.	Deposit inhibitors or Deflocculates or Emulsifying agent	Sodium salts of Sulphonic acids, salt of Phenols and carboxylic acids	It reduce foreign particles deposition on engines. It stops the passage of oil
4.	Extreme Pressure	Organic chlorine compounds, Organic sulphur and Phosphorus compounds	It forms a surface film in between two moving bodies

contd...

contd...

5. Oiliness	Vegetable oils, Castor oil, Palmitic acid, Stearic acid, Oleic acid	It increases film thickness in between metal surfaces
6. Pour point depressants	Poly alkyl benzene, Phenol	It keeps thickness constant or wax with oil
7. Thickeners	Polyester, Poly styrene	It increases viscosity
8. Viscosity index	Hexanol, Poly alkyl benzene constant at all temperatures	It keeps viscosity value
9. Anti foaming agent	Glycerol, Methyl silicone	It decreases foam formation while salts in oil

5.3.8 Semi Solid Lubricant (Greases)

Semi solid lubricants are used when oil lubricant is not suitable. Vehicle bearing and most of the places are lubricated by grease. Grease is a semi-solid lubricant. It is prepared by the following methods:

Product: Grease

Name of the Process: Saponification

Reagent: Alkali (NaOH, Ca $(OH)_2$, KOH, etc.)

Reactant: Fats or Fatty acids or Glycerides

Example,

Palmitic acid $H_3C-(CH_2)_4-COOH$

Stearic acid $H_3C-(CH_2)_7-CH=CH-(CH_2)_7-COOH$

Oleic acid $H_3C-(CH_2)_7-CH=CH-(CH_2)_7-COOH$

Reaction

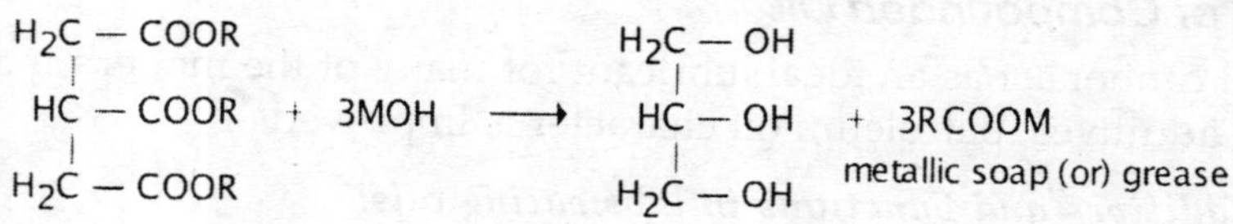

M – Na, K, Ca

Based on the metal used for saponification such metallic soaps are prepared.

For an example

M – OH: M = Na (Sodium soap)

M = Li (Lithium soap)

M = K (Potassium soap)

Alkalis are called thickening or gelling agent.

Properties of Grease

Grease contain saturated acids, possess greater thermal stability and higher dropping point. Operating temperature is lower for grease and Viscosity is higher. It forms a thick layer between moving surfaces. It reduces friction and it absorbs heat energy.

Dropping Point

The particular temperature at which semi-solid state is converted to liquid state is called dropping point of grease.

Consistency of Grease is expressed in terms of penetration

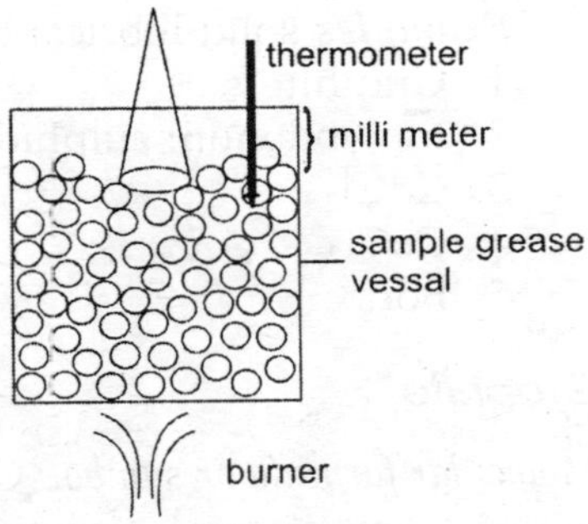

Fig. 5.6

Procedure

The distance in millimeter that a standard cone penetrates vertically into the sample under standard conditions of load, temperature and time.

Different Metallic soaps, properties and uses:

No.	Type of grease	Properties	Uses
1.	Aluminum soap grease	i. It is expensive ii. Water resistant iii. Highly sticky nature	Used for lubrication of chains and oscillating surfaces
2.	Axle grease (Resin grease, talc and mica added as fillers)	i. Very cheap ii. Water resistant	It is used in instruments which operates at low speed and high load
3.	Barium soap grease	i. Water resistant ii. It is forming sticky layer on surface	Automotive lubrication
4.	Calcium soap (Lime–base grease, cup grease)	i. It is cheapest ii. It operates beyond 70°C iii. Lesser drop point	It is used for general purpose and water pump
5.	Lithium soap grease	i. Highly water resistant ii. Heat absorber iii. Expensive	It is used in engines at low temperature
6.	Sodium soap grease	i. Slightly soluble in water ii. It has a higher droping point iii. It can be used at dry conditions	It is used in high temperature ball bearing

5.3.9 Solid Lubricants

Oils and greases cannot be used in high temperatures and high–load conditions, e.g, Aerospace devices. That is where solid lubricants are used.

Properties of Solid Lubricant

1. It has a higher dropping point
2. It can be used at high temperatures and high loads
3. It has shear strength, chemical inertness and good thermal conductivity
4. It may be either inorganic or organic compounds
5. It does not allow the external particles on the surface

Examples solid lubricants:

1. Graphite
2. Molybdenum sulphide
3. $CdCl_2$
4. $CoCl_2$
5. Borax

Graphite

Molecular formula or symbol: C (sp^2 hybridised)

Structure: Sheet structure

Type of Lubricant: Solid lubricant

Mechanism: Thin film or Boundary mechanism

Diagram

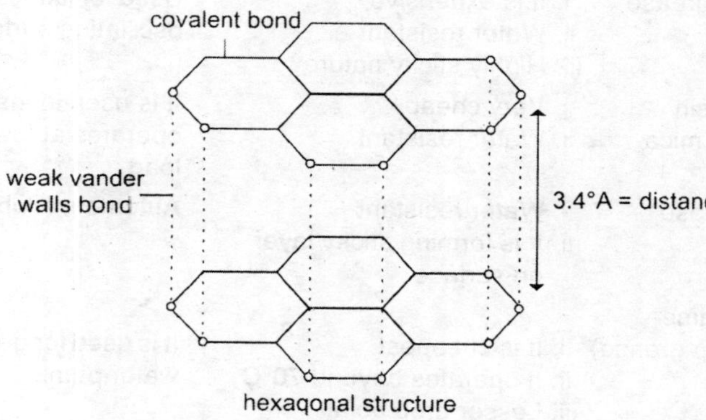

Fig. 5.7

Description

Graphite has a hexagonal sheet arrangement. Sheet layers are of one atom thickness. In between two sheet structures, van der Waals bonds are present (weak bond). The distance between two layers is 3.4°A. The layers can slide easily one over the other with little friction. So it possesses a low coefficient of friction. Graphite can be used either as powder or as colloidal dispersion in water or in oil or as grease.

Deflocculated Acheson Graphite (DAG)

1. Colloidal dispersion of graphite in water or oil is called DAG
2. Colloidal made by water is called aqua DAG
3. Colloidal made by oil is called oil DAG
4. It is ineffective in vacuum and above 370°C.

Uses

1. It is used in internal combustion engines (diesel engine), air compressors, lathes, machine shop works, etc.
2. It is also used in track joints, open gears, chains, etc.

Molybdenum disulphide
Molecular formula: MoS_2
Structure: Sandwich like structure
Type of Lubricant: Solid Lubricant
Mechanism: Boundary or Thin film mechanism

Description

Molybdenum disulphide has a sandwich structure. Sulphur layers are present between two molybdenum layers. Atoms are bonded by strong covalent bonds. Layers are bonded by weak van der Waals bonds. Molybdenum layers and sulphur layers slide over one another. It has a very low coefficient of friction. It can also be used as powder or colloidal dispersion. It is oxidized at 800°C and above.

Uses

1. It can be used under vacuum
2. In heavy machines like Military tanks, MoS_2 is used as lubricant
3. It is largely used in airframe lubrication.

Diagram

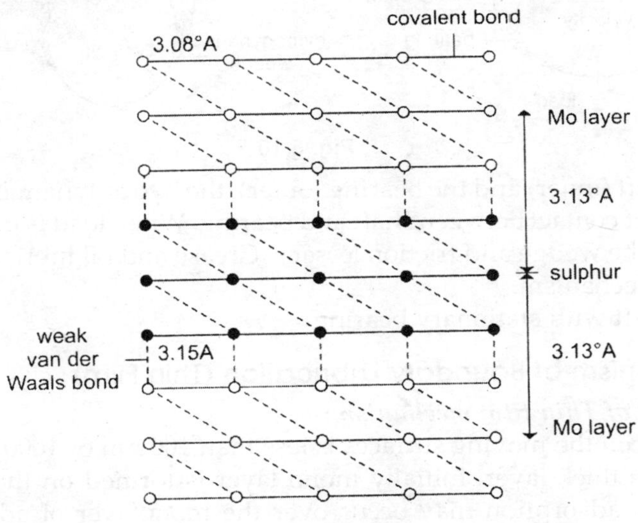

Fig. 5.8

5.3.10 Gaseous Lubricants

Gases have also been used in recent times. These are used in spindles, compressor, fans, etc.

5.3.11 Mechanism of Lubrication

Mechanism of lubrication based is on the surface action. Mechanism is classified into three types:

1. Fluid – Film Lubrication or Hydrodynamic or Thick film mechanism
2. Thin Film or Boundary Lubrication
3. Extreme Pressure Lubrication

5.3.12 Fluid–Film Lubrication or Hydrodynamic or Thick Film Mechanism

Salient features of this mechanism

In this mechanism, Thickness of the lubricant layer is not less than 100 nm or 1000°A. Coefficient of friction is very low. The coefficient value is between 0.001 to 0.03. In this mechanism, two surfaces do not contact each other directly. This reduces wear and tear. Solid–Solid interface is converted to Solid–Liquid interface

Diagram

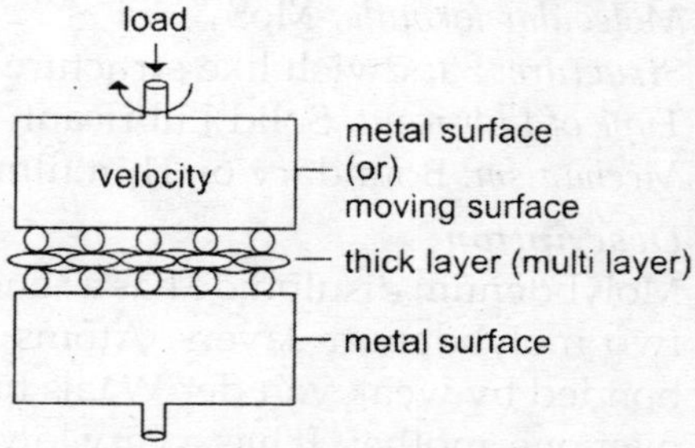

Fig. 5.9

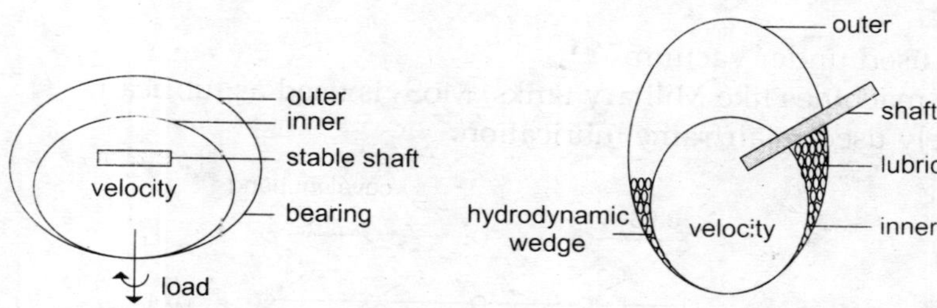

Fig. 5.10

Between the shaft (inner) and the bearing (outer), the hydrodynamic wedge is formed. There is no direct contact between shaft and bearing. When load is increased, lubricant will flowout of the wedge and friction lessens. Grease and oil lubrication are based on the thick film mechanism.

Example, Shaft with stationary bearing.

5.3.13 Mechanism of Boundary Lubrication (Thin Film)

Salient features of Thin film mechanism:

Thickness between the moving surfaces is less than 100 nm or 1000°A. It consists of a 2 to 3 molecules thick layer. Initially mono layer is formed on the moving surface. Then multilayer adsorption may occur over the monolayer of adsorbed oil. Metal surface increases, the oil layers becomes less oriented or randomly arranged. The coefficient of friction is between 0.05 to 0.15. It depends upon the oiliness of lubricant. Graphite and Molybdenum disulphide serve as good boundary lubricants

Diagram

Example, Oleic acid ($C_{17}H_{33}COOH$), Stearic acid ($C_{17}H_{35}COOH$)

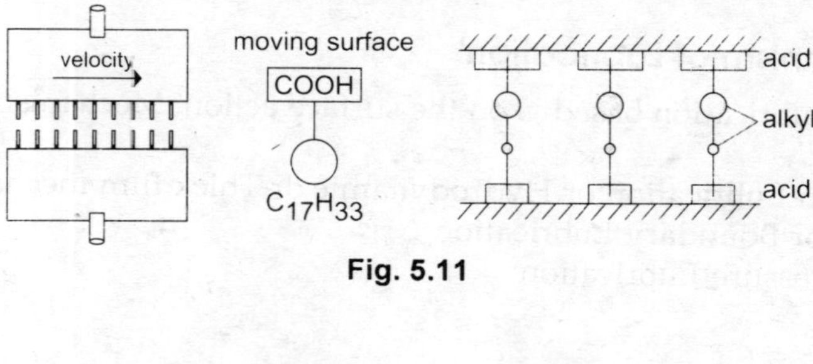

Fig. 5.11

5.3.14 Extreme Pressure Lubrication

Salient features of this mechanism

Under heavy load and high speed operating conditions, a special class of lubricants called high pressure lubricants are to be employed. Due to the frictional heat, liquid lubricants fail to stick on the surface. To avoid this, some additives are added to lubricants. These are called extreme pressure additives. Examples of such additives are organic phosphorous, sulphur and chloro compounds. After additives are added, it acts as a good lubricant under high pressure. It is ineffective on chemically inert metal surfaces (e.g. Silver, Chromium and Titanium).

5.3.15 Properties of Lubricants

1. Viscosity
2. Viscosity index
3. Cloud and pour point
4. Flash point and Fire point
5. Oiliness
6. Carbon residue
7. Aniline point

5.3.16 Viscosity

Definition: Viscosity is a measure of the internal resistance of a liquid during its flow. Unit of viscosity is centipoises or centistokes.

Apparatus used: Redwood viscometer or Say bolt viscometer.

Red wood Method: 50 ml of liquid is passed through orifice tube. This time taken for flow is mentioned by Redwood seconds.

Say bolt Method: 60 ml of liquid is passed through orifice tube. This time taken is called saybols seconds.

Orifice Tube Diameter – 3.8 mm, length – 15 mm.

Objectives

- A good lubricant must have medium viscosity
- If viscosity is high, efficiency decreases, friction increases
- If viscosity is low, it will not act as lubricant

5.3.17 Viscosity Index

1. The relation between viscosity and temperature can be explained by viscosity index.
2. Viscosity is inversely proportional to temperature, with every one degree rise in temperature, viscosity decreases (temperature rises between 100°F to 210°F).

$$\text{Viscosity} \propto \frac{1}{\text{temperature}}$$

$$(100°F = 38°C; 210°F = 99°C)$$

Viscosity Curve

The rate of change in viscosity with temperature is called viscosity temperature curve.

Procedure to Determine Viscosity Index

Initially two standard lubricant oils are taken for analysis. Std. 1 is lower viscosity oil (Naphthenic) – Gulf Coast oil (viscosity index value 0). Std. 2 is higher viscosity oil (Paraffin) – Pennsylvanian oil (Viscosity index value 100). Then unknown test lubricant should be taken whose viscosity index value is unknown. Standard Gulf coast oil is heated in red wood or saybolt Viscometer. Then gradual decrease in flow time or viscosity is noted. The same procedure followed for Pennsylvanian oil and unknown. Plot the viscosity values of three different oils versus temperature rise on a graph.

Gulf coast oil Viscosity index value is low *"L"*

Pennsylvanian oil Viscosity index value is high *"H"*

Unknown test oil Viscosity index value is unknown *"U"*

Graph

Calculation:

$$\text{Viscosity} \propto \frac{1}{\text{viscosity index}}$$

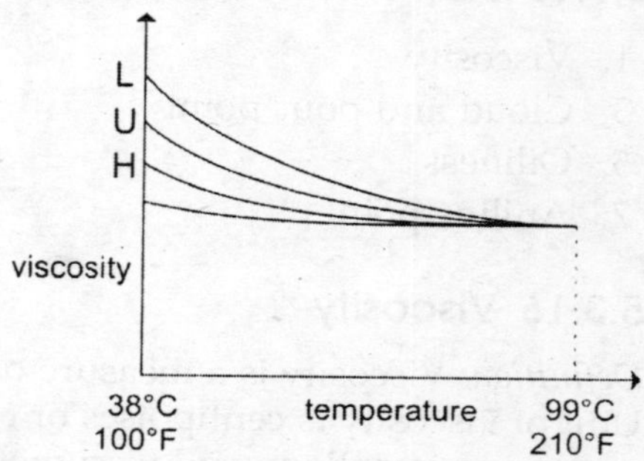

Fig. 5.12

Formula:

$$\text{Viscosity Index} = \frac{L - U}{L - H} \times 100$$

(Gulf oil) L is viscosity of low V.I oil at 38°C

(Test) U is viscosity of unknown V.I oil at 38°C

(Pennsylvanian) H is viscosity of high V.I oil at 38°C

Good lubricant oil should have minimum change in viscosity and it should have high viscosity index value. Viscosity index value increases by adding additives like, linear polymer, n-hexanol, polyisobutylene.

5.3.18 Cloud and Pour Point

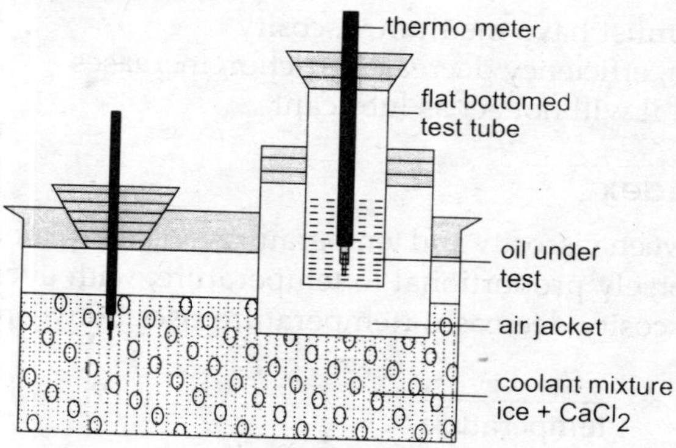

Fig. 5.13

Cloud Point (Cooling Process)

During cooling process, the particular temperature at which lubricant oil becomes cloudy or forms semi solid is called the cloud point.

Pour Point (Cooling Process)

The particular temperature at which oil does not flow and becomes solid is called pour point.

oil $\xrightarrow{\text{cooling}}$ cloud point $\xrightarrow{\text{cooling}}$ solidify $\xrightarrow{\text{cooling}}$ no longer flow or so

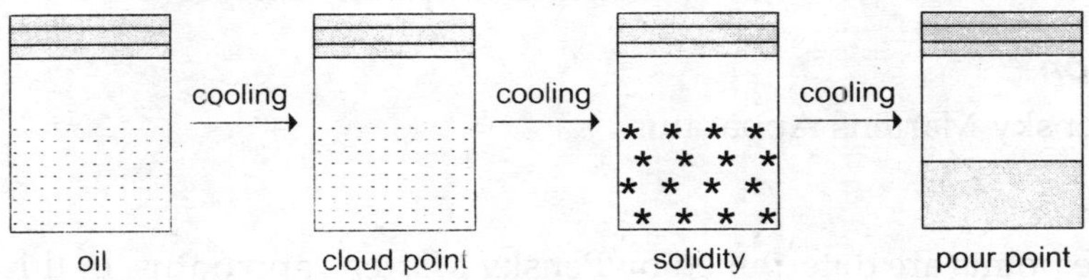

Fig. 5.14

Objectives

- Lubricant oils contain dissolved wax and other impurities
- When the oil is cooled, these impurities start crystallizing out. It gives a cloudy appearance
- Further cooling of the oil causes the oil to crystallize out and minute crystals of wax get interlocked preventing the flow of oil. This is the reason for considering the pour point.

Determination of Cloud Point and Pour Point

The apparatus consists of a flat-bottomed tube. It is of 3 cm diameter and 12 cm height. It is enclosed in an air jacket. The total jacket with tube is kept in the freezing mixture. The tube is half–filled with oil. A thermometer is inserted into the oil. Freezing mixture of ice and Calcium chloride ($CaCl_2$) is used to reduce the temperature. Temperature fall slowly inside the apparatus. The particular temperature at which cloudiness forms is noted as the cloud point. The cooling is continued and the tube is taken out for every 3°C fall of temperature and tilted. The temperature at which the oil does not flow in the tube. This temperature is called pour point. A good lubricant should have low cloud point and pour point.

Pour Point Reducing Methods

i. De-waxing,
ii. Adding pour point depressant
Example, Para flow (Commercial name) – Poly alkyl benzene

5.3.19 Flash Point or Fire Point

These two parameters were originally devised to assess the volatility of oils, which affects the consumption of oil in an IC engine.

Flash Point: The particular lowest temperature at which oil gives enough vapours to come in contact with a test flame. Kerosene flash point is 73°F

Fire Point: Fire point is the lowest temperature at which vapour of the oil burns continuously for 5 seconds, when a small flame is near it. Fire point is between 5 – 40°C higher than flash point.

Objectives

- A good lubricating oil should have higher Fire and Flash point
- It should be greater than the operating temperature
- Paraffinic oils have higher flash points than naphthenic oils

Determination

Apparatus: Pensky Martens Apparatus

Procedure

Flash and fire points are determined by Pensky Martens apparatus. In this apparatus, lubricating oil is heated in a closed cup. The lid is a movable one. It can be opened and closed. The test flame is applied near to the lid. The particular temperature at which vapour gets fire is noted in a thermometer. The temperature is called flash point. The particular temperature at which vapour continuously fire is noted as the fire point of the oil. A low flash point is indicative of evaporation losses.

Diagram

Oiliness

It is an important property of a lubricant. Oiliness is the capacity to stick on moving surfaces.

Oiliness can increase by adding vegetable oils. it should be high for Extreme–pressure lubricants.

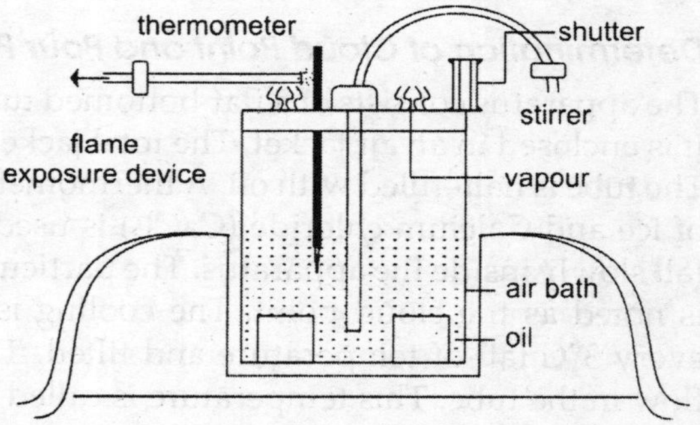

Fig. 5.15: Pensky marten apparatus

Carbon Residue

Lubricant oils tend to deposit in the combustion chamber of IC engines. Good lubricant should deposit the very least amount of carbon.

Determination

A known weight of lubricant oil is allowed to evaporate and pyrolyse.The carbonaceous residue left is expressed in percentage (1%) by weight of the sample.

Volatility

Volatility of oil can be determined by using a vapourimeter. A good lubricant has a low volatility.

Aniline Point

It explains the aromatic content in lubricant oil. Aniline point is a tendency of the lubricant to mix with aniline. A good lubricant should have a lower aniline point.

5.3.20 Choice of Lubricants

The operating conditions will detect the choice of a lubricant

No.	Operating Conditions	Types of Oils
1.	Low speed, light load	Low viscosity oil
2.	High speed, light load	Low viscosity oil
3.	Low temperature	High fluidity
4.	High temperature	Low volatility and oxidation resistance

5.4 NANO MATERIALS

5.4.1 Introduction

Nano technology can otherwise be called "nanotech". Nanotech is the study of the control of matter on an atomic and molecular scale. Nano technology deals with structures of sizes 100 nanometers or smaller and involves developing materials or devices within that size. Nano technology has the potential to create many new materials and devices with a wide range of applications, such as in medicine, electronics and energy production. The term nano technology was defined by Tokyo Science University Professor "Norio Taniguchi", In a 1974 paper as follows, "Nano Technology mainly consists of the processing of separation, consolidation and deformation of materials by one atom or by one molecule."

Fundamental Concepts

1. 1 nanometer is one billionth or 10^{-9} meter
2. Carbon — Carbon bond lengths are in the range of 0.12–0.15 nm
3. DNA double helix has a diameter of around 2 nm

Nano Science: The science which deals with the principle and properties of the materials at atomic and molecular scale in nanometers, is called nano science.

Nano Particles: The nano particles are lower in size compared with the bulk and higher in size compared with the molecule. The size ranges from 1 – 50 nm. When it is obtained as colloidal and has a tendency to remain a single crystal, it is called nano crystal. Nano particles exist naturally in the world through photochemical and Volcanic activities.

Nano Crystals: The colloidal nanomaterial is called nano crystal. It possesses electronic, magnetic and optical properties. Electronic behavior of nano crystal is called quantum dots.

5.4.2 Types of Nano Material

Nano materials are classified into three types based on dimension:

Nano Materials in

One dimension—Layers of Nano Thin Films

Two dimensions—Nano Tubes or Nano Wires

Three dimensions—Precipitates or Quantum Dots

Nano Technology: Using technology the nano materials are developed, prepared, and characterized. This is called nano technology.

Nano Chemistry: The chemistry that deals with the nano chemical material properties and its constituent's reactivity is called nano chemistry.

Example of nano material is Carbon.

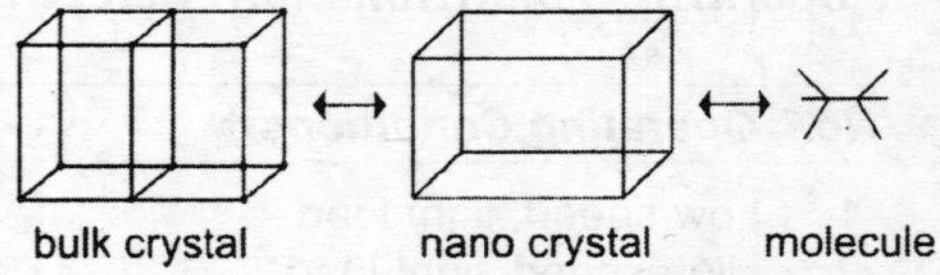

Fig. 5.16

5.4.3 Carbon Nano Tubes

Carbon exists in different allotropic forms like graphite, diamond, buck minster fullerene and nano tubes.

Graphite – Allotrope of carbon, sp^2 hybridized and sheet structure. Single sheet is called graphene. Hexagonal sheets are bonded together. It conducts electricity.

Diamond – Allotrope of carbon, sp^3 hybridized and crystal structure.

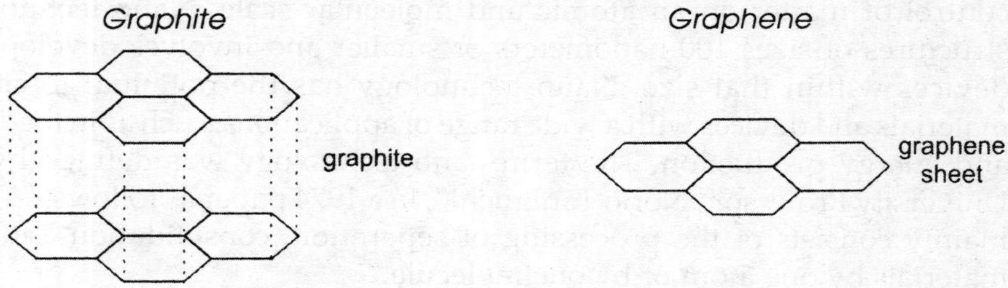

Fig. 5.17

Fullerene

It is also known as bucky ball. Buck minster fullerene has the C_{60} structure. It is a ball structure. Hexagonal structures are bonded in a ball structure.

5.4.4 Types of Carbon Nano Tubes

Carbon nano tubes (CNT) are described as a graphene sheet. Sheets are rolled into the shape of a cylinder. It is, for the reason of simplicity, called "CNT". Carbon nano tubes are large molecules of pure carbon. These are long and thin tubes with 1–3 nm diameter and 100–1000 nm lengths. Carbon atoms are linked by covalent bonds. Based on wall and rolled graphite sheet nature, there are two types. Single walled carbon nano tubes (SWCNT), diameter—1 nm multiwalled carbon nano tubes (MWCNT), diameter—10 nm.

Structure and types of single walled carbon nano tubes

Single walled carbon nano tubes (SWCNT) contain one sheet of graphite cylinders.

During rolling of grapheme sheets. It forms three types of structures.

a. Armchair
b. Zigzag
c. Chiral

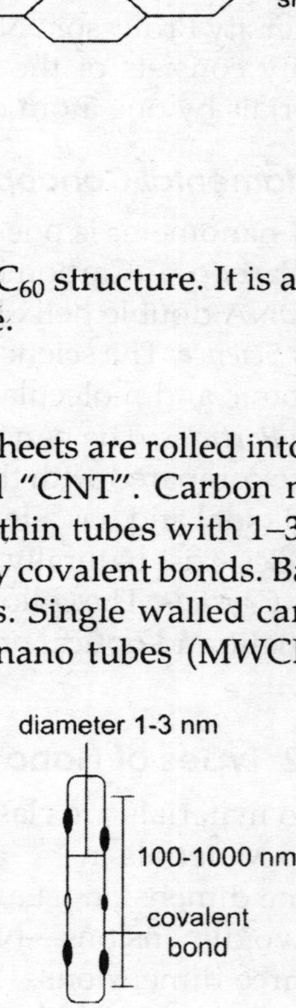

Fig. 5.18

Structure of SWCNT

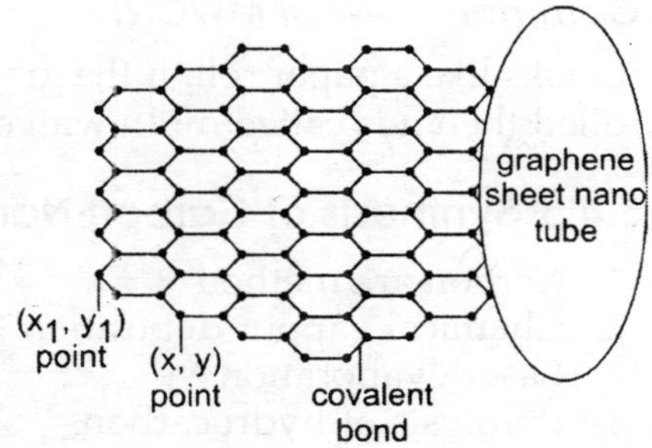

Fig. 5.19

Point is mentioned as (x, y)

- Carbon is in a hexagonal shape (Ch).
- Carbon Point is mentioned as (x_1, y_1) and (x, y)
- $\vec{j}$ is called circumferential vector. It is at right angle to the i axis
- "I" is the axis of grapheme sheet rolled up
- During the grapheme sheet being rolled at the "*i*" axis, two carbon arrangement points (x, y) and

  (x_1, y_1) $\vec{j}$ vector at right angle 90° to the i axis

Types of SWCNT

- When the "*i*" axis is parallel to the C — C bonds of the hexagons. It is called the armchair structure.
- If the "*i*" axis and points zigzag, it is called the zigzag structure.
- If the "*i*" axis and all other neighboring arrangements are very different. Then it is called the chiral structure.

Arm Chair

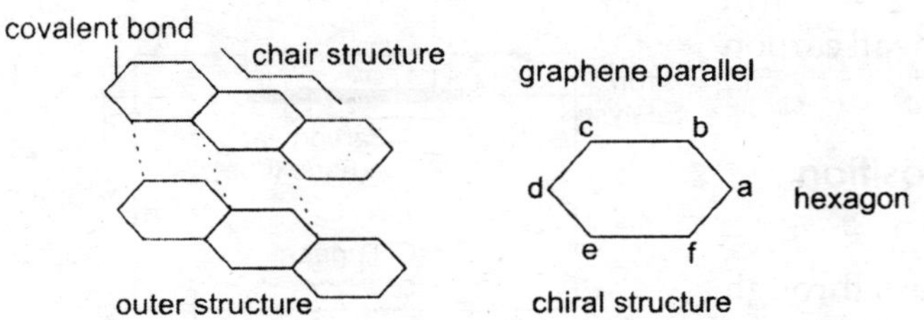

Fig. 5.20

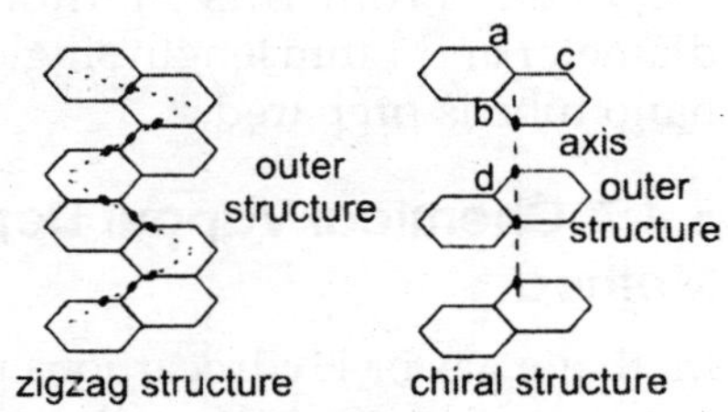

Fig. 5.21

Chiral: If any atom is bonded with four different atoms structure, then it is called chiral.

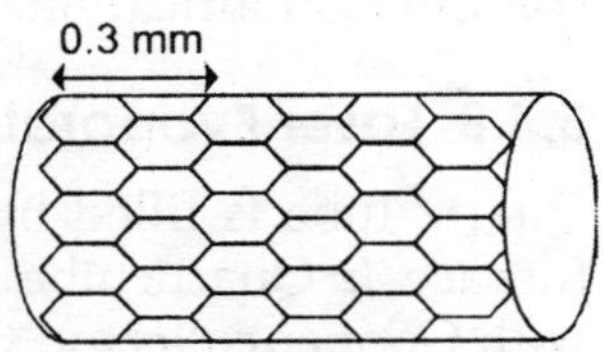

Fig. 5.22

General Structure of Carbon Nano Tubes

Carbon nano tubes are closed at both ends which involves the introduction of a pentagonal topological arrangement at each end of the tube.

Multi Wall Carbon Nano Tube

More than one grapheme sheets are rolled on the surface of the tube.

Common View of MWCNT

It looks like a paper roll. If the grapheme sheets are rolled then it is called multi wall carbon nano tube.

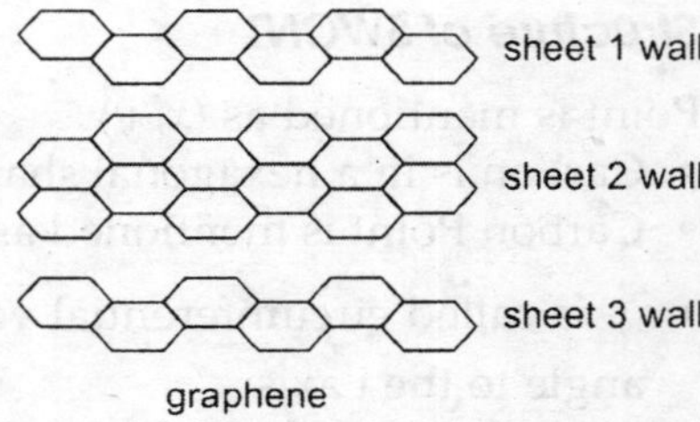

Fig. 5.23

5.4.5 Synthesis of Carbon Nano Tubes

1. Carbon arc method
2. Chemical vapour deposition
3. Laser evaporation
4. Pyrolysis of hydrocarbons

5.4.6 Carbon ARC Method

Preparation Method

Graphite electrodes are connected with an external battery. 20–25 volt external current is passed across the electrodes. Graphite electrode diameter is 2–50 μm (micrometer). Electrodes are separated by 1mm and 500 torr pressure of Helium. Carbon atom is moved from the positive end to negative end. To prepare (SWCNT), single wall carbon nano tube, Cobalt, Iron and Nickel catalysts are used. To prepare (MWCNT) multi wall carbon nano tube, catalyst should be removed. From this method, 1 – 5 nm diameter and 1 mm length single wall carbon nano tube is prepared.

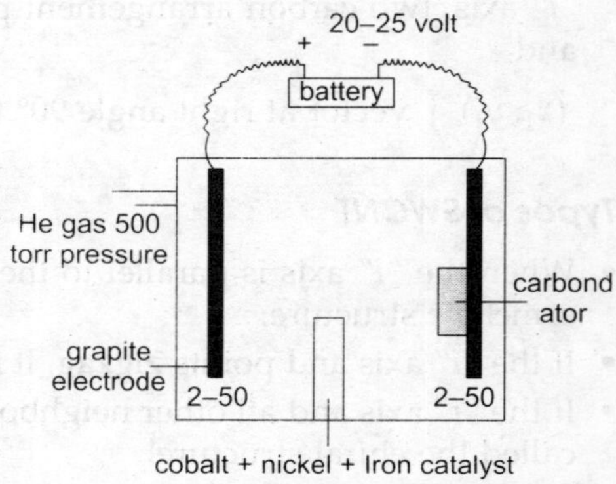

Fig. 5.24: Carbon ARC method

5.4.7 Chemical Vapour Deposition Method

Methane gas or Hydrocarbons pass through the oven. Then it is heated up to 1100°C in the absence of O_2. Gas and carbon vapour are formed. It deposit on cooling circulated Iron surface. Iron acts as a catalyst. This method produces tubes with open ends and continuous fabrication. It is a very good and cost effective method of production.

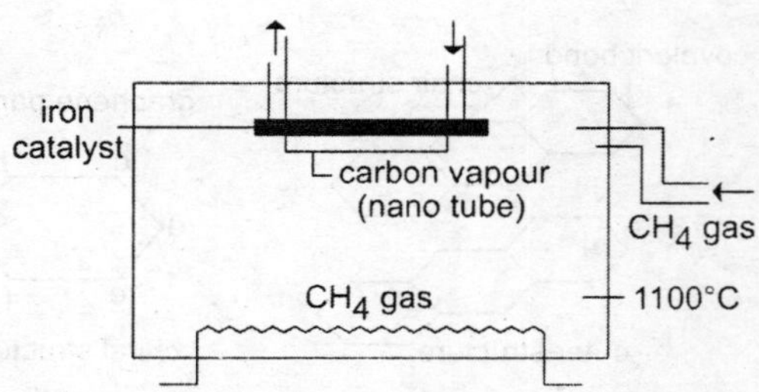

Fig. 5.25: Chemical vapour deposition method

5.4.8 Laser Evaporation

Quartz tube is filled by Argon gas and heated up to 1200°C from external electric furnace. In Quartz tube, one end is connected with laser torch and another is connected with Copper collector. Copper collector cools by cold-water circulation. At the centre of the Quartz tube, graphite target is kept along with Cobalt and Nickel catalyst. Carbon is the evaporated form of Graphite. Then the vapours are carried by a laser beam and form the tube on cold copper surfaces. By this method, 10 – 20 nm diameter and 1 '0 mm length nanotubes are prepared.

5.4.9 Pyrolysis Method

Carbon nano tubes are prepared by pyrolysis (Cracked by heating) of hydrocarbons. Example: acetylene gas pyrolysed at 700°C in the presence of Fe – Silica or Fe – Graphite catalyst under N_2 or inert conditions.

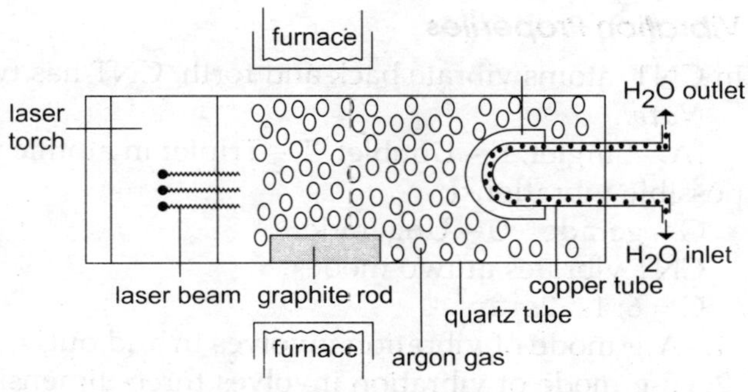

Fig. 5.26: Laser evaporation

5.4.10 Properties of Carbon Nano Tubes

1. Electrical properties
2. Kinetic property
3. Mechanical
4. Thermal
5. Vibrational properties.

Electrical Properties

- Electrical properties depend upon the diameter and chirality of the nano tube.
- Nano tubes are intermediates between metallic and semi conductor.
- Armchair acts as good conductor like metals (parallel structure).
- Zigzag and chiral structures act as semiconductors.
- Due to a defect in structure, very high electrical conductivity is created.

Kinetic Property

- SWCNT has elongation property.
- It is not affected by load. Each nanotube is having maximum load capacity.
- MWCNT are not affected by friction. There is no friction between inner and outer layers.
- Temperature does not affect the nano tube.

Mechanical Property

- Carbon nano tubes are very strong and their youngs modulus value is higher
- It has a very high load bearing capacity
- Young modulus value is 10 times greater than that of steel
- It withstands is high tension conditions.

Thermal Conductivity

- It has a very high thermal conductivity.
- Nano tube atoms are bonded by covalent bonds. Because of heat energy or thermal activity bond vibrates.
- Because of vibration, thermal conductivity is more for nano tubes.

Vibration Properties

In CNT, atoms vibrate back and forth. CNT has two normal modes of vibration.

Note:

[A – Singlet, E – Doublet, T – Triplet in atomic vibrational levels. This explains the possible vibrations]

G – gerade U – Ungerade

CNT vibrates at two modes.

C = 6; $1s^2\ 2s^2\ 2p^2$

1. A_1g mode of vibration involves in and out
2. E_2g mode of vibration involves three-dimensional oscillation. That is sphere and ellipse oscillation

5.4.11 Applications of Carbon Nano Tubes

- **Applications are being developed by using nano materials. Carbon nano tubes are used in wide ranges**
- **CNT used as Quantum Wires (Graphene):**
  Carbon nano tubes have high electrical conductivity. Instead of metallic wires, carbon nano tubes are used.
- **CNT used in memory circuits or conducting circuits:**
  In an integrated circuit, carbon nano tube composites are used. This increases shelf life of circuits and efficient integrity.
- **CNT in Microscope:**
  CNT is attached to the tip of the microscope lens. It gives clarity to the image
- **CNT in Drug Delivery Vessels:**
  CNTs are used to place the drug in a particular place. It does not affect the body any where, that's why it is used as a drug delivery vessel, like Capsules, etc.
- **CNT as Protective Shield:**
  CNTs are used to protect from electromagnetic radiation. Electronic equipments are saved from electromagnetic radiation.
- **Gas Sensor:**
  If any gas is allowed over, CNT electrical conductivity increases.
  NO_2 and NH_3 are detected by nano sensors, due to charge transfer from CNT to gas.
- **Reinforcing Material:**
  In reinforced fibers (FRP), instead of polymer fibers nano fibers or wires are used. It increases tensile strength of Fiber Reinforced Polymer (FRP), ***e.g.*** Aluminium powder is mixed with CNT (5%).
- **CNT in Storage Devices:**
  It is used in battery-like storage devices. To improve the shelf life of batteries instead of polymers, nano tubes are used. Because of its tensile strength it is used in battery, ***e.g.*** hydrogen gas stored in nano tube for fuel cell
  Nano tubes are used in lithium battery. Lithium is stored in nano tubes.

Nano Materials	Uses
Titanium dioxide	Sunscreen, food products, cosmetics
Carbon	Gecko tape
Silver	Food packaging
Zinc oxide	Surface coatings, paints
Cerium oxide	Fuel Catalyst
Silver	Bacteriostatic breaking of organic matter in waste treatment

GLOSSARY

Abrasives: Hard substances used to grind other less hard substances.

Hardness: The capacity to scratch a substance.

Natural Abrasives: Diamond and Quartz.

Refractories: The materials withstand high temperatures.

Acidic Refractory: Refractory mady by acidic chemicals.

Basic Refractory: Refractory made by basic chemicals.

Refractory Measurement: Seger Cone Method.

Lubricant: Friction reducing agent.

Dewaxing: Removal of wax from mineral oil.

Desulphurization: Removal of sulphur.

Flash Point: The lowest temperature at which oil gives enough vapours.

Nano Tube: A Very small particle tube.

Assessment Questions

Answer all questions (2 marks)

1. How are refractories classified? Give one example each.
2. Mention the characteristics or requisites of a good refractory.
3. What is meant by refractoriness of a refractory?
4. What is meant by pyrometric cone equivalent (PCE) of a refractory?
5. Name the stages in the manufacture of refractory
6. What are the important uses of refractories?
7. What is RUL? How is RUL test carried out?
8. What are the important properties of high alumina bricks?
9. Name any one of the neutral refractories and its chemical composition.
10. How is refractoriness of a refractory material determined?
11. Mention the objectives of PCE test.
12. Why silica bricks expand on heating?
13. Define porosity of a refractory.
14. What is meant by dimensional stability? Mention their types.
15. What is meant by thermal spalling? How to avoid it?

16. Name two refractories which should not be placed in direct contact with fire clay refractory. Why?
17. What are abrasives? How are they classified?
18. What are soft abrasives?
19. What is abrasive power?
20. What is hardness of an abrasive? What is its unit?
21. Explain moh's scale for different abrasives.
22. Mention some important application of abrasives.
23. Mention some important characteristics of abrasive
24. How are abrasives used?
25. What is emery?
26. How is carborundum prepared?
27. What is diamond? Mention its types.
28. How is alundum synthesized?
29. What is norbide? Mention its properties?
30. What is Garnet? Give its uses.
31. Explain the uses of alundum.
32. Define the terms "Lubricant" and "Lubrication".
33. Mention the functions of a lubricant.
34. Mention some important characteristics of a lubricant.
35. Give two examples each for liquid, semi solid and solid lubricants.
36. Give two examples of synthetic lubricating oils.
37. What are extreme pressure additives? Give examples.
38. What are oiliness carriers? Give examples or define "oiliness of a lubricant". How can this be improved?
39. What are viscosity index improvers? Give examples.
40. Define Viscosity index.
41. What is the significance of pour point of a lubricant?
42. What are pour – point depressants? Give examples.
43. What should be the flash point of a good lubricant? Give reasons.
44. Define the terms "Flash Point" and "Fire Point".
45. What are greases? How are they classified?
46. How is lithium based grease prepared? Mention the properties and uses.
47. Mention the differences between soda–base and lime-base greases.
48. What are greases? How are they prepared?
49. What are solid lubricants? Mention their applications.
50. What are the types of lubrication?
51. Under what condition extreme pressure lubricant is used.
52. What are nanomaterials?
53. What are carbon nanotubes?
54. How is carbon nanotubes classified?
55. Mention the important forms of SWNTs.
56. Name some important methods of manufacture of CNTs.
57. How are CNTs used in the fuel cell?

58. What is the main requirement for CNT containing fuel cell?
59. Give some examples for catalytic reaction carried out by using CNTs.
60. How is pyrolysis carried out?

Answer all questions (16 marks)

1. Discuss in detail about any three properties of refractories.
2. Classify refractories into different types and give one example for each type.
3. What are the significances of the three, characteristics, refractoriness, porosity and thermal conductivity of refractories?
4. What is meant by refractoriness of refractories? How is it determined?
5. What are refractories? How are they classified? Give examples for each type. Explain the terms refractoriness and thermal spalling.
6. How are silicon carbide bricks prepared? Give its important properties and uses?
7. What are garnets and carborundum? Give the properties of Boron carbide.
8. Write notes on artificial abrasives.
9. Explain the preparation and uses of silicon carbide.
10. Compare the structure of graphite and molybdenum disulphide and their uses in lubrication.
11. What are greases? When are they used? Give the preparation of Lithium grease.
12. What are solid lubricants? When are they used? Explain the structure of any one solid lubricant?
13. Write a detailed account on the graphite.
14. Indicate the preparation and uses of Lithium grease.
15. Discuss the importance of viscosity and flash point in selecting the lubricating oil for a particular use.
16. Write notes on flash and fire point of a lubricant.
17. What are carbon nanotubes? Explain their types.
18. Write synthesis, properties and application of carbon nanotubes.
19. Write notes on the followings with examples:
    i. The role of CNTs in catalysis.
    ii. CNTs in storage devices.
20. Explain the structures of CNTs.

Appendices

Appendix I

QUESTIONS AND ANSWERS

Chapter 1: Water Technology

Answer all questions (2 marks)

1. Define hardness of water.
   If water is not giving lather with soap. it is called hardness
   Reason: Some inorganic salts are present in water calcium carbonate, calcium chloride.
2. What are the salts responsible for carbonate and non-carbonate hardness of water?
   Carbonate: Calcium carbonate, magnesium carbonate.
   Non-carbonate: Calcium chloride, magnesium chloride.
3. What are the constituents that cause hardness in water?
   Carbonate: Calcium carbonate, magnesium carbonate.
   Non-carbonate: Calcium chloride, magnesium chloride.
4. Explain the terms "carbonate hardness" and "non-carbonate" hardness
   If the hardness is because of carbonate salts that is called carbonate hardness
   Example, calcium carbonate, magnesium carbonate
   If the hardness is because of non-carbonate like chloride or sulphate salts
   Example, calcium chloride, calcium sulphate.
5. Distinguish between hard and soft water.

Hard water	***Soft water***
It does not give lather with soap	It gives lather with soap
Inorganic salts are present *e.g.* $CaCO_3$, $MgCO_3$	No salts are present

6. Distinguish between carbonate hardness (ch) and non-carbonate hardness (nch)

Carbonate hardness	***Non Carbonate hardness***
If the hardness is because of carbonate salts, that is called carbonate hardness	If the hardness is because of non-carbonate like chloride or sulphate salts
Example: calcium carbonate, magnesium carbonate	***Example:*** calcium chloride, calcium sulphate
Parmanent non removable	Temporary, removable

7. How is hardness of water expressed?
   If water does not give lather with soap. It is called hardness
   Reason: Some inorganic salts present in water calcium carbonate, calcium chloride.
8. Bring out the significance of calcium carbonate equivalents?
   Molecular weight of calcium carbonate = 100
   Equivalent weight of $CaCO_3$ = 50
   To calculate hardness round figure of above values.
9. A water sample contains 73 mg $(HCO_3)_2$ per litre. Calculate the hardness in terms of $CaCO_3$.
   Amount of hardness = 73/146 × 100 = 50 mg/lit
10. Draw the structure of EDTA. what happens when EDTA is added to hard water?
    It form complex with hardness producing salts. Wine red color forms

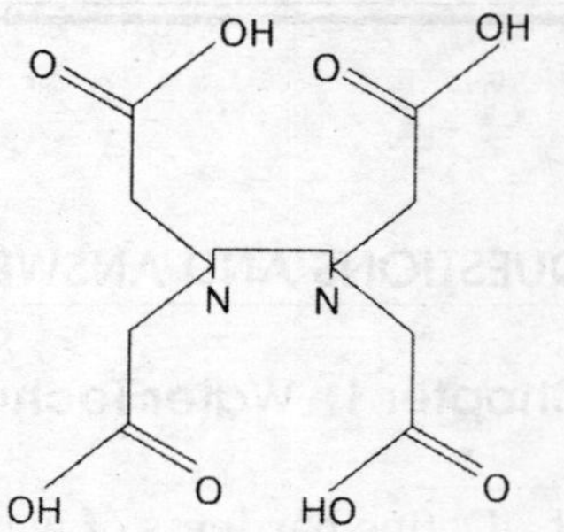

Fig. 1

11. How does eriochrome –black-T indicator function as an indicator in EDTA titration?
    It forms first complex with change of water color to wine red color.
12. Why NH_4Cl–NH_4OH buffer is used in EDTA titration.

    EBT indicator forms a complex with salts at pH of 8 to 10 that's why NH_4Cl-NH_4OH buffer is used in EDTA titration.
13. What is meant by permanent hardness of water? Mention the salts responsible for the permanent hardness of water.
    When water does not give lather with soap, it is called hardness
    Reason: Some inorganic salts present in water, e.g. calcium carbonate, calcium chloride.
14. Define alkalinity.
    The amount of acid required to neutralize the solution is called alkalinity.
    Alkalinity is because of OH^- ions, e.g. NaOH
15. How is alkalinity classified?
    a. Carbonate alkalinity ($CaCO_3$, $MgCO_3$)
    b. Bicarbonates alkalinity $(HCO_3)_2$
16. How is alkalinity determined?
    a. Carbonate alkalinity ($CaCO_3$, $MgCO_3$) – Phenolphthalein indicator
    b. Bicarbonates alkalinity $Ca(HCO_3)_2$ – methyl orange indicator.
17. Why is water softened before using in boiler?
    To avoid scales and sludge formation.
    To avoid caustic embrittle corrossion
18. What are scales and sludges?
    Scale: It is a hard and adherent deposit which is very difficult to remove
    Sludge: It is soft and non adherent and is easily removed once in boilers

19. What is meant by priming and foaming? How can they be prevented?
Priming: the process of production of wet stream is called priming.
Using treated water, good boiler designing.
Foaming: the formation of stable bubbles over the surface of the water in a boiler is called foaming. Adding castor oil, adding coagulants (small impurity is converted to big size impurity) e.g. sodium aluminate
20. What is meant by caustic embitterment? How is it prevented?
The boiler is affected by highly basic compounds like sodium hydroxide, magnesium hydroxide. Because of this, boiler gets cracked. This is called caustic embrittlement.
Adding tannin, sodium phosphate.
21. Indicate the reasons for boiler corrosion
   a. Hardness producing salts
   b. Highly basic compounds
   c. Dissolved oxygen, carbon di oxide
22. What are the requisites of drinking and boiler feed water.
PH = 6.5 – 7, chloride content = 250 ppm, hardness producing salts = 500 ppm.
23. Define the advantage of softening water. How is it carried out?
When water gives lather with soap, it is called soft water. The soft water is used as boiler water. Scale, sludges are not formed. Using sodium phosphate,and coagulants we can prepare soft water.
24. Soft water is not D M water whereas DM water is soft water, justify.
In soft water some anions are present so this is not called de-mineralized water. But demineralized water does not contain any ions so it is a soft water better than soft water.
25. What are the advantages of ion exchange process?
By this process all type of ions are removed.
It produces soft water
The water can be subjected to boilers.
26. How is exhausted resin regenerated ion exchange process?
Anion exchange resins are regenerated by adding sodium hydroxide base
Cation exchange resins are regenerated by adding hydrochloric acid
27. Give some examples for cation exchange resin.
Styrene di vinyl copolymer substituted by sulphonic acid is an example for cation exchange resin
28. Give some examples for anion exchange resin.
Styrene di vinyl copolymer substituted by amine formaldehyde is an example for anion exchange resin
29. How is water demineralised in an ion-exchanger?
The water is passed through the ion exchange resin filled column and resins are adsorbed on their surface. The outlet water is called demineralised water.

30. Give the description of sand filter.
This is the process used to filter the unwanted impurities which are bigger in size. Sand filters are used in the process of sewage water treatment. Different sizes of sands are kept in layers.
31. What is aeration of water? mention its purpose
Passing oxygen gas through the sewage water is called aeration process. To increase the COD, BOD aeration process is carried out.
32. Explain the function of a coagulant with examples.
Coagulants are the chemical compounds which form big size compounds with small size impurities. Because of this coagulating agent, impurities settle down, e.g. sodium aluminate
33. What is carbonate conditioning? For what type of boilers it is used?
Using sodium carbonate compounds permanent hardness is removed. This process is called carbonate conditioning of water. This method is used in low pressure boilers.
34. What is the role of phosphates in the internal treatment of water?
Sodium phosphate is used to remove the permanent hardness producing salt, this is called phosphate conditioning. This changes scale to sludge.
35. What are boiler troubles? Mention two different boiler troubles and their actions.
Scale formation
Sludges formation
Caustic embrittlement
Scales and sludges increase the boiling point of water and subsequently break the boilers.
36. What are calgon conditioning? How does it function in water treatment?
Calgon is sodium hexa Meta phosphate. $Na_2 [Na_4 (PO_3)_6]$
It is used to remove permanent hardness producing salts and precipitating them.
37. Mention requisites of potable water.
It should not be turbid
It should have pleasant taste
It should be colorless and odorless
38. What are the various stages in the treatment of water for domestic supply?
Raw water—screening—aeration—sedimentation—coagulation—filtration—disinfection
39. Write briefly on disinfection of water by UV treatment
Ultraviolet radiation destroys harmful bacterias. By this process no chemicals are added but it is a costly method.
40. Explain the functions of bleaching powder as a germicide
Bleaching powder is calcium hypochloride ($CaOCl_2$). This is a solid chlorinating agent. While reacting with water, chlorine gas is evolved. This will kill the harmful bacteria.
41. Mention the requisites of water used for potable water
pH, colorless, odourless,turbidless and clear. These should give the quality of water. If the above are removed then it will be used for boiler feeding and drinking purpose.

42. What is meant by "reverse osmosis"? How is it applied in the desalination of water?
When two solutions of different concentrations are separated by a semi-permeable Membrane, solvent flows from a region of lower concentration to higher concentration.
This process is called osmotic pressure. Using this method, we can remove the salts which is causes alkalinity.
43. Define the term of breaking point chlorination.
To remove the infection maters chlorine gas is passed through the water. While passing the gas initial gas is not involved in removing process. After one particular point only chlorine gas involved in process at which particular point excess chlorine gas present water is called break point of chlorination.
44. Name some of membranes employed in reverse osmosis process. State the advantages of process.
Semi permeable membrane some cellulose materials are used for reverse osmosis. by this process we can remove the salinity present in water or desalination process.
45. What are the advantages of reverse osmosis method?
    a. It is removing ionic, nonionic, colloidal, high molecular organic maters.
    b. Lifetime of the membrane is high.
46. What is blow-down operation?
Blow down operation is process of removing a portion of concentrated water by fresh water
Frequently from the boiler during steam preparation.
47. What are the disadvantages of ion exchange process
The equipment is costly.
Chemicals, setup, resin cost is very high.
48. Define desalination.
Desalinisation is the process of removing common salt from water.
The water containing dissolved salt is called brackish water.
Desalinisation can be carried out by reverse osmosis or electrodialysis.
49. What are the disadvantages of using ozone in disinfection of water?
This method is highly expensive
Ozone gas stability is less. So storage problem occurring.
50. Name the method which separate both ionic and non-ionic impurities from water.
Using reverse osmosis process we can separate the both ionic and non-ionic impurities are removed.

Answer all questions (16 marks)

1. What is meant by carbonate and non-carbonate hardness of water? Explain with example

Definition for carbonate hardness	(2 marks)
Definition for non carbonate hardness	(2 marks)
Example reactions	(5 marks)
Removal of carbonate hardness	(3 marks)
Removal of non carbonate hardness	(4 marks)

2. What is the principle of EDTA method? Describe the estimation of hardness of water by EDTA method.

Hardness estimation by EDTA method:

Aim

To estimate the amount of total, temporary and permanent hardness in the given sample of hard water. A standard solution of $CaCl_2$ is provided such that 1 ml of solution is equal to 1 mg of $CaCO_3$

Principle

The estimation is based on the complexometric titration.

a. Total hardness of water is estimated by titrating it against EDTA using EBT indicator.

   EBT + Mn + →[EBT–M] (complex)

   EBT–M (unstable complex) + EDTA → [EDTA-M] (stable complex) +EBT

   (Wine red) (Steel blue)

   EBT indicator forms wine red coloured complex with hardness causing metal ions present in water. On addition of EDTA, metal ions preferably form complexes with EDTA and steel blue EBT indicator is set free. Therefore, change of colour from wine red to steel blue denotes the end point.

b. Temporary hardness is removed by boiling the water.

   $Ca(HCO_3)_2 \rightarrow CaCO_3\downarrow + CO_2\uparrow + H_2O$

   $Mg(HCO_3)_2 \rightarrow Mg(OH)_2\downarrow + 2CO_2\uparrow$

The precipitate is filtered and the remaining permanent hardness is estimated using EDTA.

Pipette out 20 ml of standard $CaCl_2$ into a 250 ml conical flask (Standard hard water is prepared by dissolving 1 g of calcium carbonate in one litre of distilled water). Add 5 ml of buffer solution and 3 drops of eriochrome black T indicator. Titrate the solution with EDTA from the burette until the colour changes from wine red to steel blue at the end point. Repeat the titration for concordant values. Let the titre value be V_1ml.

Titration – II

Determination of total hardness

Pipette out 20 ml of sample hard water into a clean conical flask. Add 5 ml of buffer solution and 4–5 drop of eriochrome black T indicator. Titrate the wine red coloured solution with EDTA from the burette until the colour steel blue appears at the end point. Repeat the titration for concordant values. Let the titre value be V_2ml.

Titration – III

Determination of permanent hardness

Take 250 ml of hard water sample in a 250 ml beaker and boil gently for about 20 minutes. Cool, filter it into a 250 ml standard flask and make the volume up to the mark. Take 20 ml of this solution and proceed it in the same way as in titration (I). The volume of EDTA consumed corresponds to the permanent hardness of the water sample. Let the titre value be V_3 ml. Temporary hardness is calculated by subtracting permanent hardness from total hardness.

Step 1: Arriving at standard equation

1 ml of Std $CaCl_2$ = 1 mg of $CaCO_3$ (given)

V_1 ml of EDTA = 20 ml of Std. $CaCl_2$

1 ml of EDTA = $(20/V_1)$ ml of Std. $CaCl_2$

Therefore, 1 ml of EDTA = $(20 / V_1)$ mg of $CaCO_3$ (standard equation.)

Step 2: Calculation of total hardness

Volume of EDTA consumed (V_2) = ... ml

20 ml of sample hard water = V_2 ml of EDTA

1000 ml of sample hard water = $V_2 \times (1000/20)$ ml of EDTA

As per standard equation, 100 ml of given hard water = $V_2 \times (1000/20) \times (20/V_1)$ mg of $CaCO_3$

= $1000 \times (V_2/V_1)$ mg of $CaCO_3$

Therefore, total hardness = ... ppm

Step 3: Calculation of permanent hardness

Volume of EDTA consumed (V_3) = ... ml

20 ml of boiled water = V_3 ml of EDTA

1000 ml of boiled water = $V_3 \times (1000/20)$ ml of EDTA

As per standard equation,

1000 ml of given hard water = $V_3 \times (1000/20) \times (20/V_1)$ mg of $CaCO_3$

= $1000 \times (V_3/V_1)$ mg of $CaCO_3$

Therefore, permanent hardness = ... ppm

Step 4: Calculation of temporary hardness

Temporary hardness of the given sample of water = Total hardness – Permanent hardness

= ... ppm

3. Describe the principle and method involved in the determination of different types and amount of alkalinity of water.

   AIM: To determine the type and amounts of alkalinity in the given water sample. A standard solution of NaOH of strength — N is given.

   Principle

   Alkalinity in water is due to the presence of soluble hydroxides, bicarbonates and carbonates. Alkalinity can be determined by:

   - Potentiometric methods
   - Using pH meter
   - Titrimetry using different indicators

Determination of various types and amounts of alkalinity is easily carried out by titration with standard HCl employing the indicators phenolphthalein and methyl orange independently or in succession.

The following reactions occur when different types of alkalinity are neutralized with acid.

$OH^- + H^+ \rightarrow H_2O$ completed at pH 8.2–9.0 ... 1

$CO_3^{2-} + H^+ \rightarrow HCO_3^-$ ... 2

$HCO_3^- + H^+ \rightarrow (H_2CO_3) \rightarrow H_2O + CO_2$, completed at pH 4.2–5.5 ... 3

Neutralisation (1) and (2) will be notified by phenolphthalein end-point while all the three will be accounted by methyl orange end-point. Bicarbonate in eqn (3) may be due to the existence of soluble free bicarbonate salts or bicarbonates resulting from half neutralization of soluble carbonates (eqn (2))

Various steps to be followed

A known volume of water sample is titrated against std. HCl using first phenolphthalein indicator till end-point (P) and the titration is continued without break using methyl orange indicator till the equivalence end-point (M).

From the magnitudes of the P and M, the nature of alkalinity can be arrived as follows:

i. $P = M \Rightarrow$ Presence of only OH^-

ii. $2P = M \Rightarrow$ Presence of only CO_3^{2-}

iii. $P = 0, M \# 0 \Rightarrow$ Presence of only HCO_3^-

iv. $2P > M \Rightarrow$ Presence of OH^- and CO_3^{2-}

v. $2P < M \Rightarrow$ Presence of HCO_3^- and CO_3^{2-}

(Mixture of OH^- and HCO_3^- are not listed since they do not exist together and are considered equivalent to CO_3^{2-}).

PROCEDURE

Titration – I

Standardisation of HCL

Exactly 20 ml of the given standard NaOH solution is pipetted out into a clean conical flask and 2 drops of phenolphthalein indicator is added. The solution is titrated against the given HCl taken in the burette. The pink colour of the solution in the conical flask disappears at the end-point. The titre value is noted down from the burette and the titration is repeated to get concordant value.

Titration – II

Estimation of Alkalinity in Water Sample

Exactly 20 ml of water sample is pipetted out into a clean conical flask. Few drops of phenolphthalein indicator are added and titrated against the standardized HCl taken in the burette. The end-point is the disappearance of pink colour, which is noted as P. Into the same solution few drops of methyl orange indicator is added. The solution changes to yellow. The titration is continued further by adding same HCl without break till the end-point is reached. The end point is the colour change from yellow to reddish orange. The titre value is noted as M.

The experiment is repeated to get concordant values. From the magnitudes of P and M values, the type of alkalinity present in the water sample is inferred and the individual amounts are calculated and reported.

4. Describe briefly the different steps in the purification of water for drinking purposes. What is the usage of breakpoint chlorination?
Raw water—screening—aeration—sedimentation—coagulation—filtration-disinfection *(12 marks)*
Break point of chlorination with diagram *(4 marks)*

Stages in the purification process

i. *Screening*
It is the process of removing floating materials.

ii. *Aeration*
The process of mixing water with air is known as aeration. The main purpose of aeration is
a. To remove dissolved gases
b. To remove ferrous and manganous salts

iii. *Sedimentation*
It is the process of removing suspended impurities.

iv. *Coagulation*
It is the method to remove finely divided clay, silica. In this method coagulant like alum is added to water.

v. *Filtration*
It is the process of removing bacteria, colour, taste, odour, suspended particle from water using filter beds.

vi. *Sterilization or disinfection*
The process of destroying harmful bacterias is knows as sterilization or disinfection. The following methods are employed for sterilization
a. Boiling
b. Using Ozone
c. Using UV radiation
d. Chlorination

5. Define the term desalination with neat diagram, describe desalination by reverse osmosis method

Desalinisation of brackish water

a. Desalinistion is the process of removing common salt from water.
b. The water containing dissolved salt is called brackish water.
c. Desalinisation can be carried out by reverse osmosis or electrodialysis

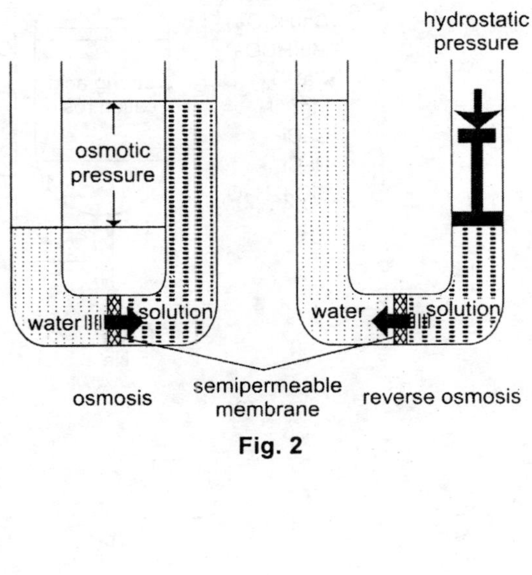

Fig. 2

Reverse Osmosis

When two solutions of different concentrations are separated by a semi- permeable membrane, solvent flows from a region of lower concentration to higher concentration. This process is called osmotic pressure.

If hydrostatic pressure in excess of osmotic pressure is applied on the higher concentration side, the solvent flows from higher concentration to lower concentration. This process is called reverse osmosis.

6. Describe de-mineralization process of water softening explain the reaction involved

External treatment

Ion exchange or demineralization

a. This process is carried out by using ion exchange resins, which are ling chain, cross linked, soluble organic polymers.
   Cation exchangers (RH_2)
b. Resins containing acidic functional groups are capable of exchanging their H-+ ions with other cations of hard water.
   Anion exchangers ($R(OH)_2$)
c. Resins containing basic functional groups are capable of exchanging their anions with other anions of hard water.

Process

d. The hard water first passed through cation exchange column.
   $RH_2 + CaCl_2 \rightarrow 2HCl + RCa$
e. The cation free water is then passed through anion exchange column.
   $R(OH)_2 + 2HCl \rightarrow RCl_2 + 2H_2O$

Regeneration

f. Cation exchanger can be regenerated by passing dil HCl or dil sulphuric acid
g. Anion exchanger can be regenerated by passing dil NaOH.

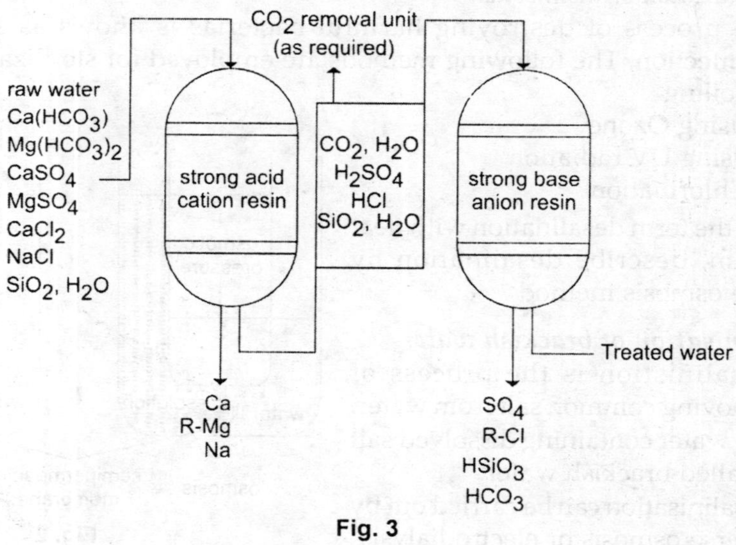

Fig. 3

7. How is internal treatment of boiler water carried out using phosphate and calgon?

Internal treatment

a. Carbonate conditioning
   - Scale formation can be avoided by adding Na_2CO_3 to the boiler water.
   - It is used only in low pressure boilers.
   - The scale forming salt like $CaSO_4$ is converted into $CaCO_3$.

$$CaSO_4 + Na_2CO_3 \rightarrow CaCO_3 + Na_2SO_4$$

b. Phosphate conditioning
   - Sodium phosphate is used in high pressures.
   - Ca and Mg reacts with phosphate as given below

$$3CaSO_4 + 2Na_2PO_4 \rightarrow Ca_3(PO_4)_2 + 3Na_2SO_4$$

Types of phosphate

a. Trisodium phosphate – used for too acidic water
b. Disodium hydrogen phosphate - used for weakly acidic water
c. Sodium dihydrogen phosphate - used for alkaline water
d. Calgon conditioning
   - Calgon is hexa meta phosphate $Na_2[Na_4(PO_3)_6]$.
   - This reacts with calcium ions forming a highly soluble complex and thus prevents scale formation.

$$2CaSO_4 + Na_2[Na_4(PO_3)_6] \rightarrow Na_2[Ca_2(PO_3)_6] + Na_2SO_4$$

With simple diagrams for each process

8. Discuss the causes and prevention of priming and foaming.

Disadvantages of using hard water in boilers

Priming and foaming

a. Due to rapid boiling of water in boilers, some liquid droplets are carried along with steam.
b. The liquid droplets carry some dissolved salts and suspended impurities along with them.
c. This phenomenon occurs due to priming and foaming.

Priming: It is the process of production of wet steam.

Control measures

a. Keeping water level lower
b. Using treated water

Foaming: It is the formation of stable bubbles above the surface of water.

Control measures

a. Adding coagulants
b. Adding anti foaming agents

9. What is the principle of reverse osmosis? How is it used for desalination?

Principle: The higher concentration solution passed through semi permeable membrane solvent passed from higher concentration to lower concentration is called reverse osmosis

Desalinisation is the process of removing common salt from water.

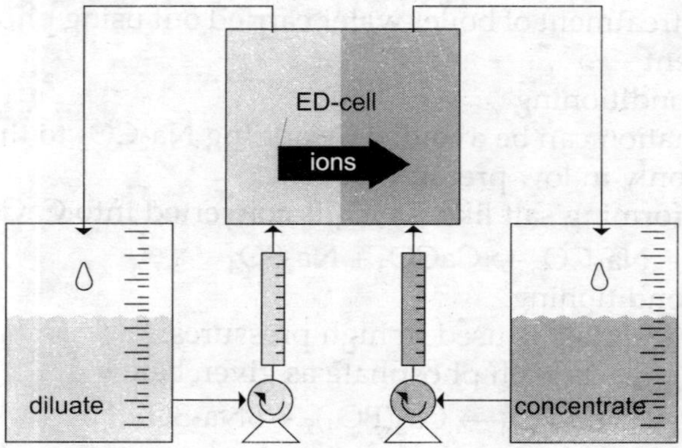

Fig. 4

The water containing dissolved salt is called brackish water.

Desalinisation can be carried out by reverse osmosis or electrodialysis.

10. What are scales and sludges? Describe the disadvantages of scale and sludge formation

Disadvantages of using hard water in boilers

a. Scale and sludge formation
b. When water is continuously converted into steam in boilers, the concentration of dissolved salts in water.
c. When the concentration of salts reach their saturation point, they are thrown out in the form of precipitate on the inner walls of the boilers.
d. Loose and slimy precipitate is called sludge. Hard coating is called scale.

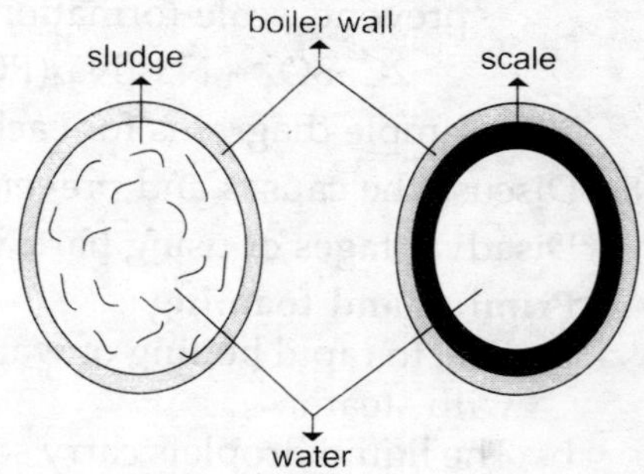

Fig. 5

Coagulation - for flocculation

Coagulant aids also known as polyelectrolytes - to improve coagulation and for thicker floc formation

Sedimentation - for solids separation, that is, removal of suspended solids trapped in the floc

Filtration - for removal of carried over floc

Disinfection - for killing bacteria

There is no unique solution (selection of processes) for any type of water. Also, it is difficult to standardise the solution in the form of processes for water from different sources. Treatability studies for each source of water in different seasons need to be carried out to arrive at most appropriate processes.

The above mentioned technologies are well developed and generalised designs are available which are used by many water utilities (public or private). In addition to the generalised solutions, a number of private companies provide solutions by patenting their technologies

11. What are the various methods by which disinfection of domestic water is carried out. Explain

UV Disinfection of water

a. The disinfection process

b. For more details on this topic, see ultraviolet.

c. Ultraviolet disinfection of water consists of a purely physical, chemical-free process. UV-C radiation in particular, with a wavelength in the 240 to 280 nanometers range, attacks the vital DNA of the bacteria directly. The radiation initiates a photochemical reaction that destroys the genetic information contained in the DNA. The bacteria lose their reproductive capability and are destroyed. Even parasites such as Cryptosporidia or Giardia, which are extremely resistant to chemical disinfectants, are efficiently reduced. UV can also be used to remove chlorine and chloramaine species from water ; this process is called photolysis, and requires a higher dose than normal disinfection. The sterilized microorganisms are not removed from the water. UV disinfection does not remove dissolved organics, inorganic compounds or particles in the water. However, UV-oxidation processes can be used to simultaneously destroy trace chemical contaminants and provide high-level disinfection, such as the world's largest Indirect Potable Reuse plant in Orange County, California.

d. The technology

e. The UV units for water treatment consist of a specialized low pressure mercury vapor lamp that produces ultraviolet radiation at 254 nm, or medium pressure UV lamps that produce a polychromatic output from 200 nm to visible and infra red energy. The optimal wavelengths for disinfection are close to 260 nm. Medium Pressure lamps are approximately 12% efficient, whilst Amalgam low pressure lamps can be up to 40% efficient. The UV lamp never contacts the water, it is either housed in a quartz glass sleeve inside the water chamber or mounted external to the water which flows through the transparent UV tube. It is mounted so that water can pass through a flow chamber, and UV rays are admitted and absorbed into the stream.

f. Inactivation of microorganisms

g. The degree of inactivation by ultraviolet radiation is directly related to the UV dose applied to the water. The dosage, a product of UV light intensity and exposure time, is measured in microwatt per second per square centimeter. Dosages for a 90% kill of most bacteria and virus range from 2,000 to 8,000 microwatt seconds per centimeter square. Dosage for larger parasites like *Cryptosporidium* require a lower dose for inactivation. As a result, the US EPA has accepted UV disinfection as a method for drinking water plants to obtain *Cryptosporidium, Giardia* or virus inactivation credits. For example, for 1-log reduction of Cryptosporidium, a minimum dose of 2,500 microwatt seconds per centimeter square is required based on the US EPA UV Guidance Manual published in 2006.

h. Weaknesses and strengths

i. Advantages

j. UV water treatment devices can be used for well water and surface water disinfection. UV treatment compares favorably with other water disinfection systems in terms of cost, labor and the need for technically trained personnel for operation: deep tube wells fitted with hand pumps, while perhaps the simplest to operate, require expensive drilling rigs, are immobile sources, and often produce hard water that is found distasteful. Chlorine disinfection treats larger organisms and offers residual disinfection, but these systems are expensive because they need a special operator training and a steady supply of a potentially hazardous material. Finally, boiling water over a biomass cook stove is the most reliable treatment method but it demands labor, and imposes a high economic cost. UV treatment is rapid and, in terms of primary energy use, approximately 20,000 times more efficient than boiling.
k. Drawbacks
l. UV disinfection is most effective for treating a high clarity purified reverse osmosis distilled water. As a matter of facts, suspended particles
m. Chloramination
n. Chloramination is the treatment of drinking water with a chloramine disinfectant. Both chlorine and small amounts of ammonia are added to the water one at a time which react together to form chloramine (also called combined chlorine), a long lasting disinfectant. Chloramine disinfection is sometimes used in large distribution systems.
o. Use
p. The maintenance of what is called a "residual" of disinfectant that stays in the water distribution system while it is delivered to people's homes is required by the Environmental Protection Agency (EPA).
q. The EPA regulations give two choices for disinfectant residual — chlorine or chloramine. Many major water agencies are changing to chloramine to better meet current and anticipated federal drinking water regulations and to protect the public health.
r. Chlorine vs Chloramine

12. Write a note on desalination by reverse osmosis and internal treatment of boiler water.
    See Answer to Question 5.
13. Explain phosphate and carbonate methods of internal treatment of boiler water.
    See Answer to Question 7.
14. How will you protect boiler from corrosion.
    It is due to the presence of:
    a. Dissolved oxygen
    b. Dissolved carbon dioxide
    c. Dissolved magnesium chloride
    - Corrosion reason: Some times water containing acidity that means pH is less than 7.5 and dissolved oxygen. This acid reacts with metal because of this metal oxides are formed this is dissolved in water. Thus metal boiler gets corrosion.

- Iron with acid reaction
- Iron with oxygen reaction
- Removing methods

15. Discuss in detail disadvantages of using hard water in boilers.
    a. Scales and sludge removing methods
    b. Priming and foaming removing method
    c. Corrosion
    d. Caustic embitterment
    e. Boiler will meet oxidants
16. Describe the internal conditioning of water.
    See Answer to Question 7.
17. Explain in detail the demineralization process for softening of water.
    Ion exchange or demineralization
    a. This process is carried out by using ion exchange resins, which are ling chain, cross linked, soluble organic polymers.
    Cation exchangers (RH_2)
    b. Resins containing acidic functional groups are capable of exchanging their H-+ ions with other cations of hard water.
    Anion exchangers ($R(OH)_2$)
    c. Resins containing basic functional groups are capable of exchanging their anions with other anions of hard water.
    Process
    d. The hard water first passed through cation exchange column.
    $RH_2 + CaCl_2 \rightarrow 2HCl + RCa$
    e. The cation free water is then passed through anion exchange column.
    $R(OH)_2 + 2HCl \rightarrow RCl_2 + 2H_2O$
    Regeneration
    f. Cation exchanger can be regenerated by passing dil HCl or dil sulphuric acid
    g. Anion exchanger can be regenerated by passing dil. NaOH.

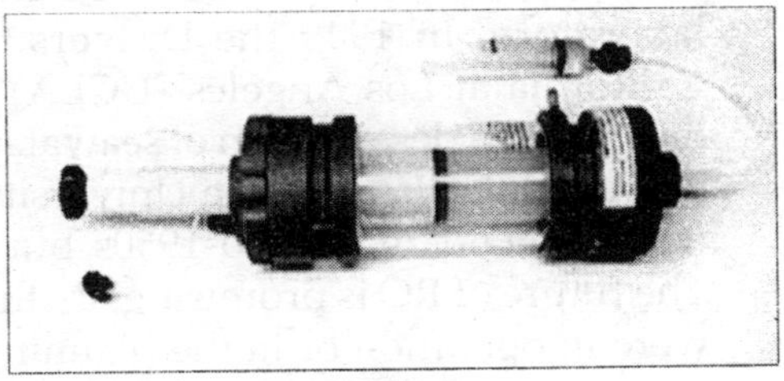

Fig. 6

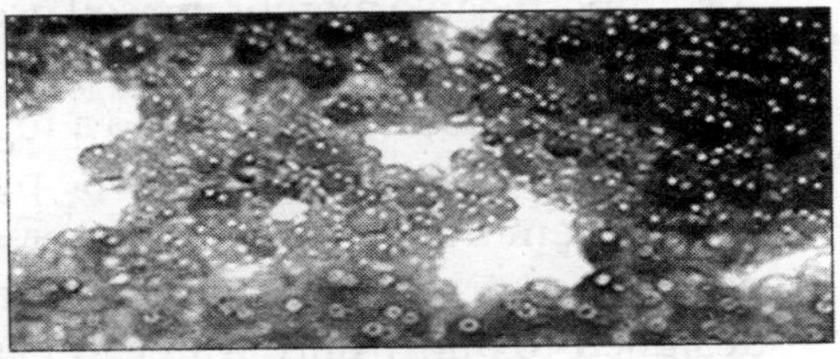

Fig. 7

18. What is meant by reverse osmosis? Explain the purification of water by reverse osomosis.
    Desalinisation of Brackish water
    Desalinisation is the process of removing common salt from water.
    The water containing dissolved salt is called brackish water.
    Desalinisation can be carried out by reverse osmosis or electrodialysis
    Reverse Osmosis
    When two solutions of different concentrations are separated by a semi-permeable membrane, solvent flows from a region of lower concentration to higher concentration. This process is called osmotic pressure.

Reverse osmosis is similar to the membrane filtration treatment process. However, there are key differences between reverse osmosis and filtration. The predominant removal mechanism in membrane filtration is straining, or size exclusion, so the process can theoretically achieve perfect exclusion of particles regardless of operational parameters such as influent pressure and concentration. RO (Reverse Osmosis), however involves a diffusive mechanism so that separation efficiency is dependent on influent solute concentration, pressure and water flux rate. It works by using pressure to force a solution through a membrane, retaining the solute on one side and allowing the pure solvent to pass to the other side. This is the reverse of the normal osmosis process, which is the natural movement of solvent from an area of low solute concentration, through a membrane, to an area of high solute concentration when no external pressure is applied.

History

The process of osmosis through semipermeable membranes was first observed in 1748 by Jean Antoine Nollet. For the following 200 years, osmosis was only a phenomenon observed in the laboratory. In 1949 the University of California at Los Angeles (UCLA) first investigated desalination of seawater using semipermeable membranes. Researches from both UCLA and the University of Florida successfully produced freshwater from seawater in the mid-1950s, but the flux was too low to be commercially viable. The future of RO is promising. By the end of 2001, about 15,200 desalination plants were in operation or in the planning stages worldwide.

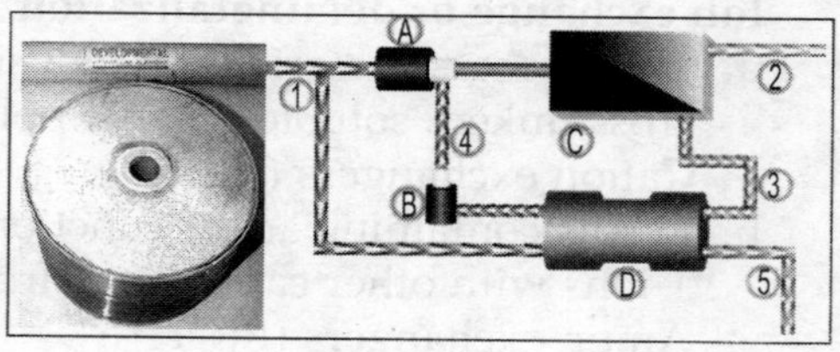

Fig. 8

Process

A semipermeable membrane coil used in desalinization. Formally, reverse osmosis is the process of forcing a solvent from a region of high solute concentration through a semipermeable membrane to a region of low solute concentration by applying a pressure in excess of the osmotic pressure.

The membranes used for reverse osmosis have a dense barrier layer in the polymer matrix where most separation occurs. In most cases the membrane is designed to allow only water to pass through this dense layer while preventing the passage of solutes (such as salt ions). This process requires that a high pressure be exerted on the high concentration side of the membrane, usually 2–17 bar (30–250 psi) for fresh and brackish water, and 40–70 bar (600–1000 psi) for seawater, which has around 24 bar (350 psi) natural osmotic pressure that must be overcome. This process is best known for its use in desalination (removing the salt from sea water to get fresh water), but since the early 1970s it has also been used to purify fresh water for medical, industrial, and domestic applications.

19. What is desalination? Name the different method of desalination. Explain any one in detail.

Desalinisation is the process of removing common salt from water.

The water containing dissolved salt is called brackish water.

Desalinisation can be carried out by reverse osmosis or electrodialysis

Reverse Osmosis

When two solutions of different concentrations are separated by a semi-permeable membrane, solvent flows from a region of lower concentration to higher concentration. This process is called osmotic pressure.

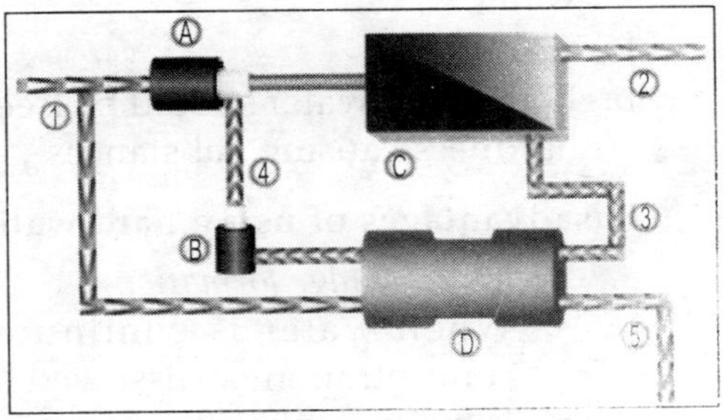

Fig. 9: Process of desalination

Reverse osmosis is similar to the membrane filtration treatment process. However, there are key differences between reverse osmosis and filtration. The predominant removal mechanism in membrane filtration is straining, or size exclusion, so the process can theoretically achieve perfect exclusion of particles regardless of operational parameters such as influent pressure and concentration. RO (Reverse Osmosis), however involves a diffusive mechanism so that separation efficiency is dependent on influent solute concentration, pressure and water flux rate. It works by using pressure to force a solution through a membrane, retaining the solute on one side and allowing the pure solvent to pass to the other side. This is the reverse of the normal osmosis process, which is the natural movement of solvent from an area of low solute concentration, through a membrane, to an area of high solute concentration when no external pressure is applied.

The process of osmosis through semipermeable membranes was first observed in 1748 by Jean Antoine Nollet. For the following 200 years, osmosis was only a phenomenon observed in the laboratory. In 1949 the University of California at Los Angeles (UCLA) first investigated desalination of seawater using semi-permeable membranes. Researches from both UCLA and the University of Florida successfully produced freshwater from seawater in the mid-1950s, but the flux was too low to be commercially viable. The future of RO is promising. By the end of 2001, about 15,200 desalination plants were in operation or in the planning stages worldwide.

20. Explain the determinations of various alkalinity of water sample.
    See Answer to Question 3.
21. Write notes on
    Carbonate and phosphate conditioning.
    See Answer to Question 7.

    Types of hardness

    Types: temporary, permanent hardness

    Temporary: carbonate hardness

    Permanent: non-carbonate hardness with chemical formula

22. What is boiler feed water? Explain the various boiler troubles.

Boiler feed water

a. The water fed into the boiler for the production of steam is called boiler feed water.

Requirements

b. Boiler feed water should be free from turbidity, oil, dissolved gases, alkali and hardness causing substances.

Disadvantages of using hard water in boilers (See also Fig. 5).

a. *Scale and sludge formation*
   - When water is continuously converted into steam in boilers, the concentration of dissolved salts in water.
   - When the concentration of salts reach their saturation point, they are thrown out in the form of precipitate on the inner walls of the boilers.
   - Loose and slimy precipitate is called sludge. Hard coating is called scale.

b. *Priming and foaming*
   - Due to rapid boiling of water in boilers, some liquid droplets are carried along with steam.
   - The liquid droplets carry some dissolved salts and suspended impurities along with them.
   - This phenomenon occurs due to priming and foaming.

   Priming: It is the process of production of wet steam.

   Control measures
   - Keeping water level lower
   - Using treated water

   Foaming: It is the formation of stable bubbles above the surface of water.

   Control measures
   - Adding coagulants
   - Adding antifoaming agents

c. *Caustic embrittlement (intercrystalline cracking)*
   - Sodium carbonate present in boiler water undergo decomposition at high pressure to give NaOH.
   - This NaOH flows into the cracks on boiler material and dissolves the surrounding area of iron as sodium ferrate.
   - Caustic embrittlement cause failure of boiler.

   Control messures
   - Using sodium phosphate
   - Adding tannin, lignin to the boiler water.

d. *Boiler corrosion*

   It is due to the presence of
   - Dissolved oxygen
   - Dissolved carbon dioxide
   - Dissolved magnesium chloride

23. Explain the estimation of hardness by EDTA method.
See Answer to Question 2.

Reverse Osmosis
Reverse osmosis is similar to the membrane filtration treatment process. However, there are key differences between reverse osmosis and filtration. The predominant removal mechanism in membrane filtration is straining, or size exclusion, so the process can theoretically achieve perfect exclusion of particles regardless of operational parameters such as influent pressure and concentration. RO (Reverse Osmosis), however involves a diffusive mechanism so that separation efficiency is dependent on influent solute concentration, pressure and water flux rate. It works by using pressure to force a solution through a membrane, retaining the solute on one side and allowing the pure solvent to pass to the other side. This is the reverse of the normal osmosis process, which is the natural movement of solvent from an area of low solute concentration, through a membrane, to an area of high solute concentration when no external pressure is applied.

24. What is portable water? What are the steps taken to obtain pure drinking water?
The raw water converted to drinking water is called portable water. It is having 14 parameters with standard values.
The following process are followed to prepare the portable water.
Desalinisation is the process of removing common salt from water.
The water containing dissolved salt is called brackish water.
Desalinisation can be carried out by reverse osmosis or electrodialysis

Reverse Osmosis
When two solutions of different concentrations are separated by a semi-permeable membrane, solvent flows from a region of lower concentration to higher concentration. This process is called osmotic pressure.

Reverse osmosis is similar to the membrane filtration treatment process. However, there are key differences between reverse osmosis and filtration. The predominant removal mechanism in membrane filtration is straining, or size exclusion, so the process can theoretically achieve perfect exclusion of particles regardless of operational parameters such as influent pressure and concentration. RO (Reverse Osmosis), however involves a diffusive mechanism so that separation efficiency is dependent on influent solute concentration, pressure and water flux rate. It works by using pressure to force a solution through a membrane, retaining the solute on one side and allowing the pure solvent to pass to the other side. This is the reverse of the normal osmosis process, which is the natural movement of solvent from an area of low solute concentration, through a membrane, to an area of high solute concentration when no external pressure is applied (see also Fig. 4).

Electrodialysis
Electrodialysis (ED) is used to transport salt ions from one solution through ion-exchange membranes to another solution under the influence of an applied electric potential difference. This is done in a configuration called an electrodialysis cell.

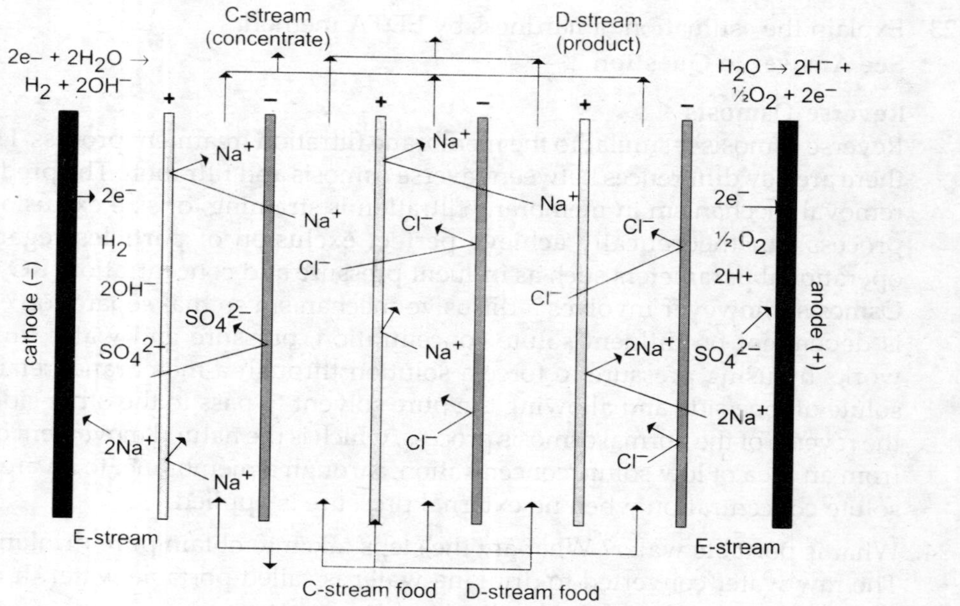

Fig. 10

The cell consists of a feed (diluate) compartment and a concentrate (brine) compartment formed by an anion exchange membrane and a cation exchange membrane placed between two electrodes. In almost all practical electrodialysis processes, multiple electrodialysis cells are arranged into a configuration called an electrodialysis.

Method

In an electrodialysis stack, the diluate (D) feed stream, brine or concentrate (C) stream, and electrode (E) stream are allowed to flow through the appropriate cell compartments formed by the ion exchange membranes. Under the influence of an electrical potential difference, the negatively charged ions (e.g., chloride) in the diluate stream migrate toward the positively charged anode. These ions pass through the positively charged anion exchange membrane, but are prevented from further migration toward the anode by the negatively charged cation exchange membrane and therefore stay in the C stream, which becomes concentrated with the anions. The positively charged species (e.g., sodium) in the D stream migrate toward the negatively charged cathode and pass through the negatively charged cation exchange membrane. These cations also stay in the C stream, prevented from further migration toward the cathode by the positively charged anion exchange membrane. As a result of the anion and cation migration, electric current flows between the cathode and anode. Only an equal number of anion and cation charge equivalents are transferred from the D stream into the C stream and so the charge balance is maintained in each stream. The overall result of the electrodialysis process is an ion concentration increase in the concentrate stream with a depletion of ions in the diluate solution feed stream

Ion exchange or demineralization

a. This process is carried out by using ion exchange resins, which are ling chain, cross linked, soluble organic polymers

Cation exchangers (RH_2)

b. Resins containing acidic functional groups are capable of exchanging their H-+ ions with other cations of hard water.

Anion exchangers ($R(OH)_2$)

c. Resins containing basic functional groups are capable of exchanging their anions with other anions of hard water.

Process

d. The hard water first passed through cation exchange column.

$RH_2 + CaCl_2 \rightarrow 2HCl + RCa$

e. The cation free water is then passed through anion exchange column.

$R(OH)_2 + 2HCl \rightarrow RCl_2 + 2H_2O$

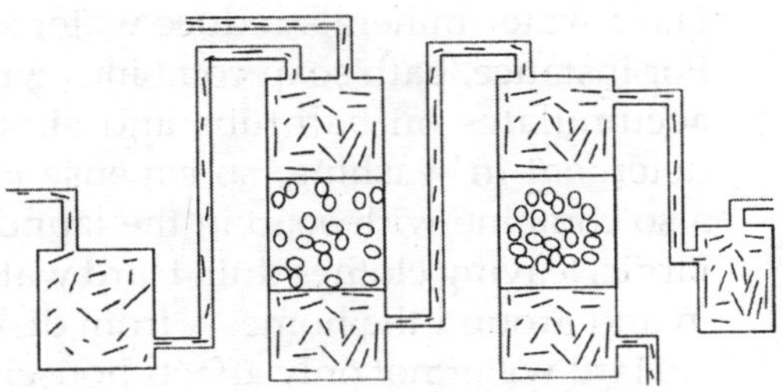

Fig. 11

Regeneration

f. Cation exchanger can be regenerated by passing dil HCl or dil slphuric acid

g. Anion exchanger can be regenerated by passing dil NaOH.

25. Explain boiler troubles? How are they caused? Suggest steps to minimize the boiler troubles.

See Answer to Question 22.

26. Explain the conventional method septic tank.

The solids that collect in the bottom of the tank and the materials that float in the scum layer are partially decomposed with the help of bacteria that occur naturally in human waste. The liquid between the solids and the scum flows out of the tank through a baffle (or a tee) and into a distribution.

Important note: As sludge accumulates in the bottom of the tank and its level rises, new wastewater coming from the house has less time for suspended particles to settle into the sludge layer. These suspended particles can flow into the absorption field. The septic tank must be pumped out periodically to remove the accumulated sludge and scum and to prevent clogging the drain-field box. The distribution box evenly separates the flow into a network of drainfield pipes. Each pipe has holes in its

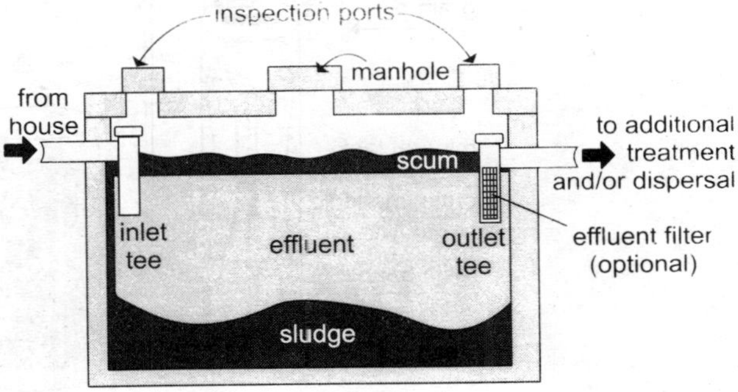

Fig. 12

underside that allow the water to drain into gravel-filled trenches. The water slowly seeps into the soil beneath the trenches where it is further treated. This process is called secondary treatment.

What makes water hard? And how does it become soft?

As water flows through layers of rock underground, it picks up loose particles and dissolves minerals from its surroundings. Because of this characteristic and the kind of rock common in many aquifers, calcium and magnesium minerals are frequently found in household water. Water with substantial amounts of calcium and magnesium is referred to as "hard water."

Hard water minerals reduce water's ability to function effectively in our homes. For instance, bath soap combines with the minerals and forms a pasty scum that accumulates on bathtubs and sinks. Homeowners must use more soap and detergent in washing, so expense for these products increases. These minerals also combine with soap in the laundry, and the residue doesnot rinse well from fabric, leaving clothes dull. Hard water spots appear on everything that is washed in and around the home — from dishes and silverware to the family car.

Hard water not only affects household cleaning, but the minerals also can build up on the inside of pipes in the plumbing system. And in water heaters, the minerals settle on the heating element, the walls of the tank, in the hot water pipes, and in faucets where they produce a scale (similar to the original rock) that reduces the efficiency and life of the hot water system.

27. Explain water softening process.

Water softeners make water work better

Water softeners combat this nuisance by eliminating the minerals that cause hard water. The most common kind of water softener is a mechanical appliance plumbed directly into the home's water supply intake (Fig. 13) The water softener exchanges calcium and magnesium with sodium in a process called ion exchange.

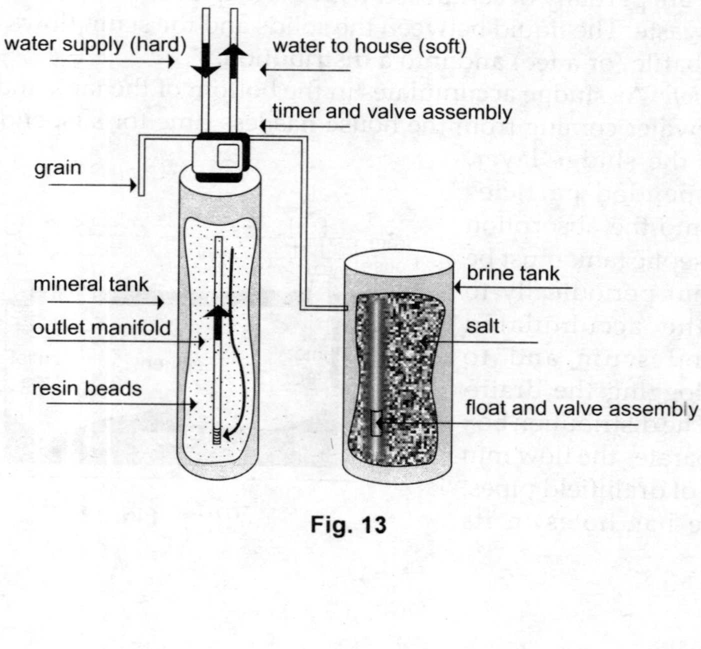

Fig. 13

The water softening system consists of a mineral tank and a brine tank. The water supply pipe is connected to the mineral tank so that water coming into the house must pass through the tank before it can be used.

The Softening Process

The normal water softening cycle operates like this: Hard water enters the mineral tank. Inside the tank, the calcium and magnesium ions carried in the water attach themselves to the beads. The surfaces of the beads eventually hold their limit of calcium and magnesium and cannot remove any more from the water. At this point the water softener must be "regenerated" (Fig. 14). The three-step regeneration cycle can be scheduled according to a timer or by a flow detection meter.

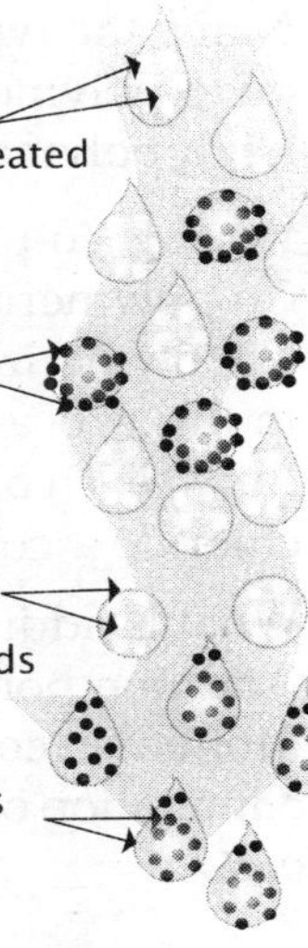

Fig. 14

Chapter 2: Polymer and Composites

Answer all questions (2 marks)

1. What is monomer?
   A polymer is a large molecule in which is formed by repeated linking of small molecules called monomers. For example, the polymer is formed by repeated linkage of simple ethylene molecules which are monomers.
   $$nCH_2 = CH_2 \longrightarrow -(CH_2 - CH_2)n-$$
   Ethylene (Monomer) Poly ethylene (polymer)
2. Define polymer.
   Polymer: Polymer is a macromolecule:
   Polymer is high molecular weight compounds formed by the combination of a large number of one (or more type of molecules of low molecular weight compounds or Polymer are macromolecules (giant molecules of higher molecular weight) formed by the repeated linking of large number of small molecules called monomer.
3. What is degree of polymerization?
   Degree of polymerization:
   The number of repeating units present in polymer chain is known as degree of polymerization. It is of two types. They are, oligo polymers and high polymers. It can be calculated by following equation:
   Dp = (molecular weight of polymer) (M) / (molecular weight of a monome) (m)
   $$4CH_2 = CH_2 \rightarrow CH_2 - CH_2 - CH_2 - CH_2 - CH_2 - CH_2 - CH_2 - CH_2$$
4. Write the steps of free radical mechanism.
   All the three mechanism said above process through three major steps namely,
   Chain initiation
   Chain propagation
   Chain terminatio

5. Name the types of degree of polymerization.
oligo polymer:
High polymer:

6. What is co-polymerisation?
Co-polymerisation:
Copolymerization is the joint polymerization of two or more different monomers. In this process no side product have been formed.
Example I: copolymerization of acrylonitrile and vinyl chloride to give poly vinyl chloride – co-acrylonitrile

7. What is addition polymerization?
Addition polymerization or chain growth polymerization:
Monomer combines to form polymer through addition reaction and without elimination of any molecules such as water, hydrochloric acid, Hydrogen bromide, etc.

Formation of polyethylene:

Ex: $n(CH_2{=}CH_2) \longrightarrow (CH_2-CH_2)n$

Formation of polyvinyl chloride:

$$n(CH_2{=}\underset{\displaystyle Cl}{\underset{|}{CH}}) \longrightarrow (CH_2-\underset{\displaystyle Cl}{\underset{|}{CH}})n$$

8. What is condensation polymerization? Condensation polymerization or stepwise polymerization:
These are formed by combination of monomers with elimination of simple molecules such as water or methanol. There are two main type of condensation polymer: polyester and polyamides.
Ex: Terylene also called Decron pol

9. Distinguish between addition and condensation polymerization.

S.No	Addition	Condensation
1	Monomers add to give a polymer and no other side product have been formed	Monomer combine to give polymer and by-products are formed.
2	Homochain polymer is obtained	Heterochain polymer is obtained
3	Thermoplastics are produced	Thermosetting plastics are produced
4	Linear chain polymer is obtained	Cross linked polymer are obtained
5	They have very low molecular weight	They have high molecular weight

10. What is initiators
Free radical which are produced by the decomposition of compound such as hydrogen peroxide benzyl peroxide, etc. these compound are called initiators

11. What is plastics?
    Plastics are high molecular weight organic material which can be moulded or formed into any stable shapes by application of heat and pressure.
12. Classification of plastics:
    Plastics can be classified based on structure and usage. They are,
    a. Commodity plastic b. Engineering plastics or high performance plastics.
13. Difference between commodity plastic and engineering plastic:

S.No	Commodity plastic	Engineering plastic
1.	Posses low abrasion resistance	Posses high abrasion resistance
2.	Posses poor dimensional stability	Posses good dimensional stability
3.	Posses low mechanical strength	Posses high mechanical strength
4.	It is general purpose plastics	It is high performance plastics
5.	It cannot be used at high temperature	It can be used at high temperature

14. Disadvantages of plastics:
    a. They undergo embrittlement at low temperature
    b. They are very high softness
    c. They undergo deformation underload
    d. They posses low heat-resistant and poor ductility
    e. They are very easily combustible
15. Difference between thermo and thermosetting plastics.
    Difference between thermoplastic and thermosetting plastic:

S.No	Thermoplastic	Thermosetting plastic
1.	They are formed by addition polymerization	They are formed by condensation polymerization
2.	Adjacent polymer chains are held together by weak vander waals force	Adjacent polymer chains are held together by strong covalent bonds called cross links
3.	They are soften on heating and becomes hard and rigid on cooling	They are hardened on heating and once set they cannot be softened again.
4.	Plasticity is reversible	Plasticity is irreversible
5.	They are weaker	They are strong
6.	They are linear polymer	They are cross linked polymer
7.	They are soluble in organic solvent	They are insoluble in organic solvent
8.	They can be remoulded	They cannot be remoulded

16. Distinguish between commodity and engineering plastics.
    Difference between commodity plastic and engineering plastic:

S.No	Commodity plastic	Engineering plastic
1.	Posses low abrasion resistance	Posses high abrasion resistance
2.	Posses poor dimensional stability	Posses good dimensional stability
3.	Posses low mechanical strength	Posses high mechanical strength

17. Mention preparation and uses of Polyvinyl chloride (PVC):

Preparation

PVC is prepared from vinyl chloride by the following step.

Step I

Acetylene react with hydrochloric acid at 60–80 pC in presence of metal chloride as catalyst. This gives vinyl chloride

$$CH=CH + HCI \longrightarrow CH_2=CHCI$$

Step II

During this process water emulsion of vinyl chloride is heated in presence of benzyl peroxide or hydrogen peroxide in an autoclave under pressure.

18. Preparation of Teflon

$$2CHCIF_2 \longrightarrow CF_2=CF_2 + 2HCI$$

$$CF_2=CF_2 \longrightarrow -(CF_2-CF_2-)n$$

19. Preparation of nylon 6, 6 and properties

Preparation

Nylon-6, 6 is obtained by the polymerization of adipic acid with hexamethylene diamine.

$$NH_2-(CH_2)_6-NH_2 + HOOC-(CH_2)_4-COOH \longrightarrow (HN-(CH_2)_4\ NH-OC(CH_2)_4-CO)n$$

Properties

- They are translucent
- They are whitish
- They are high melting polymer

Uses

- Nylon is used as a substitute for metal in bearings, gearings, etc.
- It is used for fibre which is used in making socks, dresses, carpets, etc.
- They are used for making filaments for ropes.

20. Uses of perlon – U

They are used for making industrial plugs, sockets, switches, helmets, hair drier bodies, camera, solar collectors, telephone booths, lens, photographic film, etc.

21. What is polycarbonate?

Preparation

It is prepared by interaction of diphenyl carbonate with bis phenol A 2,2 bis(4 – hydroxyphenyl) propane

Diphenyl carbonate + bis pheol A $\longrightarrow$ polycarbonate

22. How is perlon – U is prepared?

They are prepared by treating diisocyanate and diol

$$\underset{\text{hexamethylene diisocyanate}}{O=C=N-(CH_2)_6-N=C=O} + \underset{\text{1, 4 – betanediol}}{HO-(CH_2)_4-OH} \longrightarrow$$

$$\underset{\text{perlon – U}}{-\{CO-NH-(CH_2)_6-NH-CO-O-(CH_2)_4-O-\}}$$

23. What is Elastomer? Rubber or Elastomer:

    Rubber or elastomer is non-crystalline high molecular weight linear polymer having elastic properties.

24. Raw rubber cannot be used – why?
    - The rubber has poor strength
    - It has little durability
    - It has poor elasticity
    - It is easily oxidized by nitric acid and sulphuric acid

25. What is vulcanization of rubber?

    The process consists of addition of 1–3% of sulfur to the rubber compound and heat under pressure. Thus, it get excellent changes in its properties (increased elasticity, tensile strength, chemical resistance).

26. What are the characteristics of rubber

    Rubber has elasticity

    It has resilience

    It has low thermal and low electrical conductivity

    It is soluble in various organic solvents such as petrol, petroleum ether, benzene, etc.

    Vulcanized rubber has greater tensile strength, high stiffness and toughness and resistance to chemicals, oxidation process

27. What are the constituents of composites

    There are two important constituents which are matrix phase and dispersed phase.

28. What is FRP? A material system consisting of a mixture of two or more macro constituents, which are mutually insoluble in differ form or composition and forming distinct phases

29. What are the properties of FRP?
    - It posses lower densities
    - It posses higher tensile strength
    - It posses higher corrosion resistance

30. Name some important FRP.

    Glass - FRP composites:

    Aramid fibre reinforced polymer composites:

    Short fibre reinforced polymer composites:

    Long fiber reinforced polymer composites:

    Boron FRP composites:

31. Mention the application of FRP.

    It is used in the manufacture of structural components of air craft, helicopter and sports material

32. What is SBR?

    Mention some uses of SBR. SBR (Styrene – butadiene rubber / Buna –S rubber / GR – S

It is a co-polymer of about 75% butadiene and 25% styrene. The Buna – S is the first synthetic rubber developed during the time of Second World War by US in order to overcome the scarcity of natural rubber.

33. What are the characteristics of butyl rubber?
It is amorphous powder
It is soluble in hydrocarbon

34. Write the preparation of PC
Preparation: It is prepared by interaction of diphenyl carbonate with bis phenol A 2, 2 bis (4 – hydroxyphenyl) propane
Diphenyl carbonate + bis pheol A polycarbonate

35. Explain the drawbacks of raw rubber.
The rubber has poor strength
It has little durability
It has poor elasticity
It is easily oxidized by nitric acid and sulphuric acid

36. What is the aim of vulcanization
To improve certain properties of rubber like tensile strength, abrasion resistance, chemical resistance, etc. to convert the linear / branched chain rubber into cross linked structure.

37. Comparison between raw rubber and vulcanized rubber.
Comparison between raw rubber and vulcanized rubber:

S.No	***Raw rubber***	***Vulcanized rubber***
1.	Little durability	High durability
2.	Very high water adsorption capacity	Very low water adsorption capacity
3.	Tensile strength is very low	Tensile strength is high
4.	Very low operating temperature (10 – 60 pC)	Very high operating temperature (40 – 100 pC)
5.	It has high elongation	It has moderate elongation

38. Preparation of polyester resin.
Preparation:Polyester resin or condensation product of dicarboxylic acid with dihydroxyl alcohol. It is less resistance to heat and moisture.

39. Properties of PET
It is less resistance to alkalis.

40. Name some plasticizer
Dioctyl phthalate, dibutyle phthalate

41. What are the types of PVC?
Rigid PVC or Unplasticized PVC
Non-rigid PVC or plasticized PVC

42. What are thermoplastics
These are resins which soften on heating and become hard and rigid on cooling

43. What are thermosetting plastics?
These are resins which harden or gain strength on heating and once set they cannot be softened again. They cannot remoulded and reshaped. They are hard and more brittle.

44. What is corfoam?
They are used as leather substute material.

45. What is commodity and engineering plastics?
General purpose plastics have low to medium mechanical properties. They are used for the manufacture of commodity items (house hold materials). They are crystalline structure.
   1. Engineering plastics are obtained from high polymer resist. They posses high mechanical strength, roughness, environmental resistance and higher thermal resistance and mainly used engineering field.

46. What are the properties of nylon 6, 6?
Preparation:
Nylon-6,6 is obtained by the polymerization of adipic acid with hexamethylene diamine.

47. Comparison between raw rubber and vulcanized rubber.
Comparison between raw rubber and vulcanized rubber:

S.No	***Raw rubber***	***Vulcanized rubber***
1.	Little durability	High durability
2.	Very high water adsorption capacity	Very low water adsorption capacity
3.	Tensile strength is very low	Tensile strength is high
4.	Very low operating temperature (10–60 pC)	Very high operating temperature (40 – 100 pC)
5.	It has high elongation	It has moderate elongation

48. Preparation of polyester resin.
Preparation:
Polyester resin or condensation product of dicarboxylic acid with dihydroxyl alcohol.

49. Properties of PET.
   - It is less resistance to heat and moisture.
   - It is less resistance to alkalis.

50. Name some plasticizer.
Dioctyl phthalate, dibutyle phthalate
What are the types of PVC.
Rigid PVC or unplasticized PVC
Non-rigid PVC or plasticized PVC

51. What are thermoplastics?
These are resins which soften on heating and become hard and rigid on cooling

52. What are thermosetting plastics?
These are resins which harden or gain strength on heating and once set they cannot be softened again. They cannot remoulded and reshaped. They are hard and more brittle.

53. What is corfoam?
They are used as leather subsitute material

54. What is commodity and engineering plastics?
General purpose plastics have low to medium mechanical properties. They are used for the manufacture of commodity items (house hold materials). They are crystalline structure.
Engineering plastics are obtained from high polymer resist. They posses high mechanical strength, roughness, environmental resistance and higher thermal resistance and mainly used engineering field.

Answer all questions (16 marks)

1. Distinguish between addition and condensation polymerization. Difference between addition and condensation polymerization:

S.No	Addition	Condensation
1.	Monomers add to give a polymer and no other side product have been formed	Monomer combine to give polymer and by-products are formed.
2.	Homochain polymer is obtained	Heterochain polymer is obtained
3.	Thermoplastics are produced	Thermosetting plastics are produced
4.	Linear chain polymer is obtained	Cross linked polymer are obtained
5.	They have very low molecular weight	They have high molecular weight

2. What is copolymerization? Give examples

Co-polymerisation:
Copolymerization is the joint polymerization of two or more different monomers. In this process no side product have been formed.
Example I: Copolymerization of acrylonitrile and vinyl chloride to give poly vinyl chloride co-acrylonitrile.
Example II: 1,3 butadiene react with styrene to give Buna-s-rubber.

3. Define the term of addition polymerization and condensation polymerization.
Addition polymerization or chain growth polymerization:
Monomer combines to form polymer through addition reaction and without elimination of any molecules such as water, hydrochloric acid, Hydrogen bromide, etc.

Formation of polyethylene

Example, $n(CH_2=CH_2)\ (CH_2-CH_2)n$

Formation of polyvinyl chloride

$$(CH_2-\underset{\underset{Cl}{|}}{CH})n \qquad n(CH_2=\underset{\underset{Cl}{|}}{CH})$$

,I. Condensation polymerization or stepwise polymerization:

These are formed by combination of monomers with elimination of simple molecules such as water or methanol. There are two main type of condensation polymer: polyester and polyamides.

Ex: Terylene also called Decron polyester

4. Write the preparation of nylon and its uses.

preparation

Nylon-6, 6 is obtained by the polymerization of adipic acid with hexamethylene diamine.

Uses

- Nylon is used as a substitute for metal in bearings, gearings, etc.
- It is used for fibre which is used in making socks, dresses, carpets, etc.
- They are used for making filaments for ropes.
- They are used for making tooth brushes and films, tyre-cords.
- It can be sterilized with steam or boiling water, so it is suitable for medicinal applications.

5. Write preparation, properties and uses of nylon 6,6.

Preparation

Nylon-6, 6 is obtained by the polymerization of adipic acid with hexamethylene diamine.

Properties

- They are translucent
- They are whitish
- They are high melting polymer
- They posses high temperature stability
- They posses good abrasive resistance
- They are insoluble in common organic solvents

Uses

- Nylon is used as a substitute for metal in bearings, gearings, etc.
- It is used for fibre which is used in making socks, dresses, carpets, etc.
- They are used for making filaments for ropes.
- They are used for making tooth brushes and films, tyre-cords.
- It can be sterilized with steam or boiling water, so it is suitable for medicinal applications.

6. Name the engineering plastics and discuss about them Engineering plastics or high performance plastics:

Engineering plastics are obtained from high polymer resist. They posses high mechanical strength, roughness, environmental resistance and higher thermal resistance and mainly used engineering field. They are used for load bearing application.

Characteristics or properties of engineering plastics:

They posses-

- High abrasion resistance
- High mechanical strength
- High dielectric constant
- Have high load bearing capacity
- Light weight
- Resistance to weather and chemicals

7. Distinguish between thermo and thermosetting plastics.

Difference between thermoplastic and thermosetting plastic:

S.No	Thermoplastic	Thermosetting plastic
1.	They are formed by addition polymerization	They are formed by condensation polymerization
2.	Adjacent polymer chains are held together by weak vaner waals force	Adjacent polymer chains are held together by strong covalent bonds called cross links.
3.	They are soften on heating and becomes hard and rigid on cooling	They are hardened on heating and once set they cannot be softened again.
4.	Plasticity is revesible	Plasticity is irreversible
5.	They are weaker	They are strong
6.	They are linear polymer	They are cross linked polymer
7.	They are soluble in organic solvent	They are insoluble in organic solvent
8.	They can be remoulded	They cannot be remoulded

8. What are the drawbacks of rubber and how to improve it.

Drawback / Demerits of natural rubber:

- The rubber has poor strength
- It has little durability
- It has poor elasticity
- It is easily oxidized by nitric acid and sulphuric acid

Vulcanization

The process of heating the raw rubber with sulfur at 100–140 pC is called vulcanization.

Objective

- To improve certain properties of rubber like tensile strength, abrasion resistance, chemical resistance, etc.
- To convert the linear / branched chain rubber into cross linked structure.

Process of vulcanization

The process consists of addition of 1–3% of sulfur to the rubber compound and heat under pressure. Thus, it get excellent changes in its properties (increased elasticity, tensile strength, chemical resistance

Compare the raw rubber and vulcanized rubber

S.No	Raw rubber	Vulcanized rubber
1.	Little durability	High durability
2.	Very high water adsorption capacity	Very low water adsorption capacity
3.	Tensile strength is very low	Tensile strength is high
4.	Very low operating temperature (10–60 pC)	Very high cperating temperature (40–100 pC)
5.	It has high elongation	It has moderate elongation

9. Explain the properties and uses of FRP.

Properties of FRP

- It posses high yield strength, fracture strength and fatigue life.
- It posses very high mechanical properties, high corrosion resistance, heat resistance properties.
- Posses lower density
- Posses excellent corrosion resistance.
- It posses lower densities
- It posses higher tensile strength
- It posses higher corrosion resistance
- They are metal like ductile
- They have capable of absorbing energy

Uses

They are used for making structural components in aircraft and helicopter.
It is used in the manufacture of structural components of air craft, helicopter and sports material.
It is used for making automobile brakes and clutches.
It is used for making automobile parts and flouring plastic pipes.

10. Write about some engineering plastics with their special characters. Engineering plastics or high performance plastics:

Engineering plastics are obtained from high polymer resist. They posses high mechanical strength, roughness, environmental resistance and higher thermal resistance and mainly used engineering field. They are used for load bearing application.

Characteristics or properties of engineering plastics:

They posses-

- High abrasion resistance
- High mechanical strength
- High dielectric constant
- Have high load bearing capacity
- Light weight
- Resistance to weather and chemicals

Explain the composite and its types.

11. Explain about FRP

Composites

A material system consisting of a mixture of two or more macro constituents, which are mutually insoluble in differ form or composition and forming distinct phases.

Constituent of composites

There are two important constituents which are matrix phase and dispersed phase.

Matrix phase

It consists of continuous body constituent which encloses composites and gives bulk form. It is further classified into 3 types:

Metal matrix composites

- Ceramics matrix composites
- Polymer matrix composites

Dispersed phase

It consists of structural constituent which determine the internal structure of composites, e.g. fiber, particulates, flakes, whiskers.

Types of composites

Polymer composites / FRP composites

Preparation:

FRP is produced by bonding a fibre material with a resin matrix and curing them under the pressure and heat. It consist of three components such as filaments, polymer matrix and bonding agent.

Bonding material

It is used in FRP. Some of the bonding materials are glass, alumina, carbon, graphite, boron, etc.

a. Write the preparation of nylon and its uses.
b. Write preparation, properties and uses of nylon 6,6.
c. Name the engineering plastics and discuss about them.
d. Distinguish between thermo and thermosetting plastics.

What are the drawbacks of rubber and how to improve it.

Drawback / Demerits of natural rubber

- The rubber has poor strength
- It has little durability
- It has poor elasticity

e. It is easily oxidized by nitric acid and sulphuric acid

Compare the raw rubber and vulcanized rubber

Vulcanization:

The process of heating the raw rubber with sulfur at 100–140 pC is called vulcanization.

Objective:

- To improve certain properties of rubber like tensile strength, abrasion resistance, chemical resistance, etc.
- To convert the linear / branched chain rubber into cross linked structure.

Process of vulcanization

The process consists of addition of 1–3% of sulfur to the rubber compound and heat under pressure. Thus, it get excellent changes in its properties (increased elasticity, tensile strength, chemical resistance).

Uses of rubber

- It is used for the manufacture of tyres.
- It is used for preparing conveyor belts.
- It is used for manufacturing hoses.
- It is used for the manufacture of toys, sports items, etc.

Comparison between raw rubber and vulcanized rubber:

S.No	Raw rubber	Vulcanized rubber
1	Little durability	High durability
2	Very high water adsorption capacity	Very low water adsorption capacity
3	Tensile strength is very low	Tensile strength is high
4	Very low operating temperature (10–60 pC)	Very high operating temperature (40–100 pC)
5	It has high elongation	It has moderate elongation

Properties of rubber

Rubber has elasticity

- It has resilience
- It has low thermal and low electrical conductivity
- It is soluble in various organic solvents such as petrol, petroleum ether, benzene, etc.
- Vulcanized rubber has greater tensile strength, high stiffness and toughness and resistance to chemicals, oxidation process

Synthetic rubber

Preparation

SBR (Styrene – butadiene rubber / Buna –S rubber / GR – S rubber)

f. It is a co-polymer of about 75% butadiene and 25% styrene. The Buna Write the mechanism of free radical polymerization. Mechanism of addition polymerization:

Addition polymerization can be understand by the self addition or repeated addition of monomers. There are three type of mechanism by which we can produce chain polymer.

- Free radical mechanism
- Ionic mechanism
- Co-ordination mechanism

All the three mechanism said above process through three major steps namely,

- Chain initiation
- Chain propagation
- Chain termination

Free radical mechanism

The chain initiation is brought by the free radicals formed by the decomposition of compounds called initiators. Initiators are generally unstable compounds and decompose into free radicals.

Hemolytic fission

If R – R is an initiator and when energy is supplied to this molecule it split into two symmetrical compounds having one electron on each. This type of splitting is called as hemolytic fission.

g. Explain the properties and uses of FRP.

Properties

- Posses lower density
- Posses excellent corrosion resistance.
- It posses lower densities
- It posses higher tensile strength
- It posses higher corrosion resistance

It give effective reinforcement due to high aspect ratio, high surface area and toughness, strength, heating stability.

- They are metal like ductile
- They have capable of absorbing energy
- It have excellent stiffness
- It has excellent compressive strength
- It posses good abrasion resistance
- It posses good creep resistance
- It posses good dimensional stability

h. Write about some engineering plastics with their special characters.

Polycarbonates (PC) (lexan / Merlan)

Preparation

It is prepared by interaction of diphenyl carbonate with bis phenol A 2, 2 bis(4-hydroxyphenyl) propane)

Diphenyl carbonate + bis pheol A $\longrightarrow$ polycarbonate

Properties

- They are soluble in organic solvents and alkalis.
- They have stiffness
- They have good dimensional stability
- They have high impact and tensile strength
- They have transparency

Uses

They are used for making industrial plugs, sockets, switches, helmets, hair drier bodies, camera, solar collectors, telephone booths, lens, photographic film, etc.

- It is used for fibre which is used in making socks, dresses, carpets, etc.
- They are used for making filaments for ropes.
- They are used for making tooth brushes and films, tyre-cords.
- It can be sterilized with steam or boiling water, so it is suitable for medicinal applications.
- It has good resistance moisture

i. Give the preparation properties and application of polyamides.
Polyamides (Nylon – crystalline polymer)

Preparation

Nylon-6, 6 is obtained by the polymerization of adipic acid with hexamethylene diamine.

Properties

- They are translucent
- They are whitish
- They are high melting polymer
- They posses high temperature stability
- They posses good abrasive resistance
- They are insoluble in common organic solvents

Uses

- Nylon is used as a substitute for metal in bearings, gearings, etc.
- It is used for fibre which is used in making socks, dresses, carpets, etc.
- They are used for making filaments for ropes.
- They are used for making tooth brushes and films, tyre-cords.
- It can be sterilized with steam or boiling water, so it is suitable for medicinal applications.

j. Explain the composite and its types.

Composites

A material system consisting of a mixture of two or more macro constituents, which are mutually insoluble in differ form or composition and forming distinct phases.

Constituent of composites

There are two important constituents which are matrix phase and dispersed phase.

Matrix phase

It consists of continuous body constituent which encloses composites and gives bulk form. It is further classified into 3 types:

Metal matrix composites

- Ceramics matrix composites
- Polymer matrix composites

Dispersed phase

It consists of structural constituent which determine the internal structure of composites, e.g. fiber, particulates, flakes, whiskers.

Types of composites

Polymer composites/FRP composites

Preparation

FRP is produced by bonding a fibre material with a resin matrix and curing them under the pressure and heat. It consist of three components such as filaments, polymer matrix and bonding agent.

Bonding material

It is used in FRP. Some of the bonding materials are glass, alumina, carbon, graphite, boron, etc.

Filament

It is used in FRP. Some of the filament used is glass fiber, metallic fiber.

Polymer matrix

It is also used in FRP. Examples of polymer matrix are polyester, epoxy resins, phenolic resins, polyamide polymer resins.

Polyester resins

It provides very good strength and mechanical properties.

Epoxy resins

It provides good mechanical strength.

Phenolic resin

It used for the purpose of withstanding high temperature.

Properties of FRP

- It posses high yield strength, fracture strength and fatigue life.
- It posses very high mechanical properties, high corrosion resistance, heat resistance properties.

Types of FRP composites

Carbon – FRP composites / high performance composites

Properties

- Posses lower density
- Posses excellent corrosion resistance.

Uses

It is used in the manufacture of structural components of air craft, helicopter and sports material.

Glass - FRP composites

- It posses lower densities
- It posses higher tensile strength
- It posses higher corrosion resistance

Uses

It is used for making automobile parts and flouring plastic pipes.

Aramid fibre – reinforced polymer composites:

Short fibre – reinforced polymer composites:

It give effective reinforcement due to high aspect ratio, high surface area and toughness, strength, heating stability.

Uses

It is used for making automobile brakes and clutches.

Long fiber – reinforced polymer composites:

Properties

- They are metal like ductile
- They have capable of absorbing energy

Uses

They are used for making structural components in aircraft and helicopter.

Boron FRP composites

Properties

- It have excellent stiffness
- It has excellent compressive strength

Uses

It is used for the manufacture of horizontal and vertical tail in aeroplane, stiffening spares, etc.

Alumina FRP composites

Properties

- It posses good abrasion resistance
- It posses good creep resistance
- It posses good dimensional stability

Uses

It is used for the manufacture of engine parts in automobile industry and components of turbine engine.

Chapter 3: Surface Chemistry

Answer all questions (2 marks)

1. Define adsorption and adsorbate.

   Adsorption is the phenomenon of concentration of molecules of a gas or liquid at a solid surface. The substance which is held on the surface of the solid is called the adsorbate.

2. What is sorption?

   Sorption is the process in which both adsorption and absorption takes place simultaneously.

3. What is chemical adsorption or chemisorptions? Give an example chemical adsorption is the one, in which the adsorbed molecules are held on the surface of the adsorbent by chemical bonds (covalent bond or ionic bond).

   Example: Adsorption of hydrogen on nickel.

4. What is physical adsorption or physisorption? Give an example physical adsorption is the one, in which the adsorbed molecules are held on the surface of the adsorbent by weak physical or Vander Waal's forces of attraction.

   Example: Adsorption of Hydrogen or oxygen on charcoal.

5. Mention some important characteristic of adsorption.

   a. Spontaneous.

   b. Always accompanied by evolution of heat.

   c. Accompanied both by decrease in enthalpy and entropy of the system.

   d. Selective process.

   e. The rate of adsorption depends on temperature.

   f. Adsorption is a physical phenomenon, but accompanied by a chemical change.

   g. It is specific; it depends on the nature of adsorbent and adsorbate.

6. How do chemisorptions differ from physisorption?

Physisorption	chemisorption
• It is caused by intermolecular Vander Waal's forces (weak).	It is caused by chemical bond formation (Strong).
• Heat of adsorption is low (20–40 k.cal/mol).	Heat of adsorption is high (40–400 k.cal/mol).
• Reversible	Irreversible
• Adsorption decreases with increase in	It increases with temperaturetemperature.
• Multilayer adsorption occurs.	Only monolayer adsorption occurs.

7. How will you increase the activity of an adsorbent?
Effect of activation of adsorbent
Activation leads to increase in the surface.
Creation of rough surface
   a. By mechanical rubbing,
   b. By subjecting to some chemical reactions on the solid adsorbent.

   Increasing effective area of the surface
   a. By sub dividing the solid adsorbent into fine particles.
   b. By heating of solid adsorbent in superheated steam now its pores are opened and adsorption increases.
8. Explain the effect of temperature on adsorption.
   a. *Physical adsorption:* It occurs rapidly at lower temperature and decreases with increase in temperature.
   b. *Chemical adsorption:* It increases with increase in temperature and then decreases.
9. Explain the function of activated charcoal with example. An adsorbent adsorbs substances (solutes) from the solution in two ways.
   a. Solid substances adsorb dissolves substances from solutions.
      Example: Activated charcoal adsorbs coloring matter present in sugar solution.
   b. An adsorbent also adsorbs certain substance from the solution in preference to other substances.
      Example: Charcoal adsorbs non-electrolytes more readily than electrolytes from a solution.
10. What is an adsorption isotherm? Freundlich's adsorption isotherm?
Adsorption isotherm is a relationship (or a graph) between magnitudes of adsorption with pressure at constant temperature.The relationship between the magnitude of adsorption (x/m) and pressure (P) can be expressed mathematically by an empirical equation known as Freundlich adsorption isotherm, i.e.,
$x/m = KP\ 1/n$
11. What is Langmuir's adsorption isotherm?
According to Langmuir, s assumptions, when the gas molecules strike a sold surface, some of the molecules are adsorbed and some of these are desorbed. Thereby dynamic equilibrium is established between adsorption and desorption. If A is gas molecule and M is surface.

12. Derive the mathematical expression of Langmuir's adsorption isotherm.
According to Langmuir's assumptions, when the gas molecules strike a solid surface, some of the molecules are adsorbed and some of these are desorbed. Thereby dynamic equilibrium is established between adsorption and desorption. If A is gas molecule and M is surface then,

$$A_{(g)} + M\ (\text{surface}) \underset{kd}{\overset{ka}{\rightleftharpoons}} AM$$

13. What is demerit of Langmuir's adsorption isotherm?
Langmuir adsorption isotherm holds good at lower pressure but fails at high pressure.
14. What are promoters?
Promoters are defined as the substances, which increase the activity of a catalyst.
    a. Promoters change the lattice spacing
    b. Promoters increase the peaks and cracks
15. What is catalytic poisoning?
Catalytic poison is defined as a substance which destroys the activity of the catalyst to accelerate a reaction. This process is called catalytic poisoning.
16. What is the effect of temperature and pressure on the adsorption of hydrogen gas on charcoal?
Adsorption of H_2 on charcoal is rapidly at lower temperature and decreases with increase in temperature. But the rate of adsorption increases with the increase of pressure.
17. How is arsenic poisoning is removed from the body?
Colloidal ferric hydroxide is administered, which adsorbs arsenic poison and is removed from the body by vomiting
18. Define ion-exchange adsorption?
Ion-exchange adsorption is defined as, "the process of releasing the ion and adsorbing another like ion".
19. How is evaporation of water in lake minimized?
Due to scarcity of water during summer a layer of stearic acid is spread over water lakes and reservoirs. The adsorbed stearic acid on the surface of water minimizes evaporation of water.
20. Explain adsorption of demineralization of water.
The water coming out of the anion exchanger is completely free cations and anions. This water is known as demineralised water or deionised water.
21. What are the conditions of Freundlich's adsorption isotherms?
    a. Freundlich equation is purely empirical and has no theoretical basis.
    b. The equation is valid only upto a certain pressure and invalid at higher pressure.

c. The constants K and 'n' are not temperature independents, they vary with temperature.
d. Freundlich's adsorption isotherm fails, when the concentration of adsorbate is very high.

22. Explain adsorbent.

The solid that takes up a gas or a solute from the solution is called the adsorbent.

23. Explain types of adsorption.

a. Physical adsorption or Physisorption or Vander Waal's adsorption.
b. Chemical adsorption or Chemisorption or Activated adsorption.

24. Explain limitation of Freundlich's adsorption isotherm?

a. Freundlich's equation is purely empirical and has no theoretical basis.
b. The equation is valid only up to a certain pressure and invalid at higher pressure.
c. The constants K and 'n' are not temperature independents, they vary with temperature.
d. Freundlich's adsorption isotherm fails, when the concentration of adsorbate is very high.

25. Derivation of Freundlich's adsorption isotherms?

This equation may be derived from the result observed from the above graph. Thus,

a. *At low pressure:* Adsorption increases with pressure
   $x/m \propto P'$ or $x/m = KP$
b. *At high pressure:* Adsorption is almost constant
   x/m = constant or $x/m = K$
c. *At intermediate (normal) pressure:* Adsorption depends on 0 to 1 power of pressure (i.e., fraction a power of pressure)
   $x/m = P'1/n$
   or $x/m = KP1/n$ ... 1

where, n = whole number.

This equation (1) is called Freundlich's adsorption isotherm.

Taking logarithm on both sides, the above equation becomes

$\log x/m = \log K + 1/n \log P$

On plotting log x/m Vs log P, a straight line is obtained with a slope of 1/n and intercept log k.

26. Explain GAC.

Granular Activated Carbons (GAC)

These carbons have a diameter of greater than 0.1 mm and are generally used for the adsorption of gases and vapours.

27. Explain PAC.

Powdered Activated Carbons (PAC)

These carbons have a diameter of less than 200 mesh and are generally used in purification of liquids.

28. What are the properties of activated carbon?
The properties of AC are governed not only by the nature of raw materials but also by the method of activation used. Property of AC is highly helpful in specifying the carbon for a specific applications.
For Example: 1. Decolorizing activated carbon:
It is usually employed as powders. Generally raw material for this type have a weak structure.
For Example: Saw dust and lignite yield carbon of this kind.
2. Vapour – adsorbent carbon: It is used in the form of hard granules.
For Example: Coconut shells, fruit pits, briquette coal yield carbon of this kind.

29. What is adsorptive capacity?
Adsorptive Capacity: It is used to find out the effectiveness of the carbon in removing desired constituents such as COD, colour, phenol, etc. from the waste water.

30. Define phenol number?
Phenol number: It is used the find out the ability of AC to remove taste and odour compounds.

31. Define iodine number?
Iodine number: It is used to find out the ability of AC to adsorb low-molecular weight substances.

32. What are advantage of using granular activated carbon?
As the carbon adsorbs organic materials, the apparent density of the carbon particles increases and encourages migration of the heavier or spent carbon downward.

33. What are the disadvantage of using granular activated carbon. Upflow columns may have more carbon fines in the effluent than downflow columns, because upflow tends to expand, not compress, the carbon. Bed expansion allows the fines to escape through passage ways created by the expanded bed.

34. What are the applications of activated carbon?
There is no particular AC that is effective for all purposes.
    - In odour control
    - As a decolourant
    - In solution purification
    - In gas masks-to adsorb poisoning gases.
    - In Air Conditioning-to control odour.

35. What is specific action of the catalyst?
The adsorption depends on the nature of both the adsorbent (catalyst) and the adsorbate (reactants). So, different catalysts cannot possess the same affinity for the same reactants. Thus, the action of the catalyst is specific.

36. Explain the Enhanced activity of a rough surfaced catalyst
It possess " Cracks", "Peaks", "Corners", etc. and consequently have larger number of active centers. These active centers increase the rate of reaction.

37. Explain adsorption of solute from solution.
   a. Solid substances adsorb dissolves substances from solutions.
      For Example: Activated charcoal adsorbs coloring matter present in sugar solution.
   b. An adsorbent also adsorbs certain substance from the solution in preference to other substances.
      For Example: Charcoal adsorbs non-electrolytes more readily than electrolytes from a solution.
38. What are factors influencing of adsorption of solute from solution.
   a. Effect of temperature and concentration
      Negative adsorption
   b. Effect of surface area
      Surface area of the adsorbent increases, adsorption increases.
   c. The nature of the solute adsorbed
      The extent of adsorption is usually greater, when the molecular weight of the solute is high.
39. Define occlusion.
   The adsorption of gas on a solid is sometimes called occlusion.
40. Define adsorbate.
   The substance which is held on the surface of the solid is called the adsorbate.
41. Explain absorption.
   Absorption implies that a substance is uniformly distributed throughout the body of the solid or liquid. Adsorption is the surface phenomenon, but absorption is bulk phenomenon.
42. Explain fullers earth process.
   It is used in large quantities for refining petroleum and vegetable oils, which adsorb unwanted materials
43. Explain froth floation process.
   The sulphide ores (Pbs, zns,cu2s)are freed from silica and other earthy matter by froth floation process (oil adsorbs sulphide ores, only, but not other)
44. What are medical uses of ion exchange adsorption method?
   a. Excess sodium salts from the body fluids can be removed by giving a patient a suitable ion exchanger to eat.
   b. Weakly basic anion –exchanger are used to remove excess acid or "acidity" in the stomach.
45. What is the scope and study of activated carbon in pollution abatement of air and waste water?
   a. Preparation and properties of activated carbon.
   b. Use of granular and powdered activated carbon.
   c. Regeneration of activated carbon
46. Explain down flow carbon contactors.
   It usually consist of two or three columns operated in series or in parallel.
   The water or air is applied to the top of the column and withdrawn at the bottom.
   The AC is held in place with an under drain system at the bottom of the column.

Provision for back washing and surface washing is usually necessary to limit the headless build up due to the removal of particle material with the carbon column.

47. What are the advantage and disadvantage of down flow carbon contactors.?

Advantage

Adsorption of organic materials and filtration of suspended solids are accompanied in a single step.

Disadvantages

Downflow filters may require more frequent back washing because of the accumulation of suspended materials on the surface of the contactor.

Plugging of carbon pores may require premature removal of the carbon for regeneration, thereby decreasing the useful life of the carbon.

48. Explain up flow carbon contactors.

In the up flow columns, the polluted water or air moves upward from the base of the column

49. What are the advantage and disadvantage of up flow carbon contactor?

Advantage

As the carbon adsorbs organic materials, the apparent density of the carbon particles increases and encourages migration of the heavier or spent carbon downward.

Disadvantage

Upflow columns may have more carbon fines in the effluent than downflow columns, because upflow tends to expand, not compress, the carbon. Bed expansion allows the fines to escape through passage ways created by the expanded bed.

50. What is the method using powdered d activated carbon?

In this method PAC is added directly to the effluent coming out from the various biological treatment processes. In the case of biological treatment plant effluent, PAC is added to the effluent in a contacting basin. After some time, the carbon is allowed to settle at the bottom of the tank, and the treated water is then removed from the tank.

Since carbon is very fine, a coagulant such as polyelectrolyte may be added to aid the removal of the carbon particles or filtration through granular medium filters may be required.

Chapter 4: Non-Conventional Energy Sources and Storage Devices

Answer all questions (2 marks)

1. Define nuclear fission.

Nuclear fission it is the nuclear reaction in which heavy isotopes are split into lighter nuclei on bombardment by neutrons. Fission reaction of U^{235} is given below

$$_{92}U^{235} + {_0n^1} \longrightarrow {_{36}Kr^{92}} + {_{56}Ba^{141}} + 3\,{_0n^1} + \text{energy}$$

2. Mention few important characteristics of nuclear fission.
   - The enormous amount of energy released during the nuclear fission is due to the loss in some mass.
   - During nuclear fission, the sum of the masses of the products formed is slightly less than the sum of masses of target species and bombarding neutron.
   - The loss in mass gets converted into energy according to Einsteins equation $E = mc^2$
     Where, C = velocity; m = mass loss; E = energy
3. What is nuclear fusion? Give an example.
   Nuclear fusion it is the Process of combination of lighter nuclei into heavier nucleus with simultaneous liberation of large amount of energy, e.g. solar system
   $${}_1H^2 + {}_1H^2 \square\ 2He4 + \text{Energy}$$
4. What is thermo nuclear reaction?
   The Process of combination of lighter nuclei into heavier nucleus with simultaneous liberation of large amount of energy. The fusion reaction takes place at 100 million °C temperature, e.g. solar system
5. Give any two differences between nuclear fission and fusion.

S. No	nuclear fission	nuclear fusion
1.	This process gives chain reaction.	This process does not give chain reaction.
2.	It emits the radioactive rays	It does not emits any kind of the radioactive rays

6. What is nuclear chain reaction?
   The process of propagation of a fission reaction by multiplication in three at each fission is called nuclear chain reaction.
7. What is super critical mass and subcritical mass of U^{235}?
   Super critical mass: If the mass of fissionable material (U^{235}) is more than the critical mass then it is called super critical mass. Subcritical mass: If the mass of fissionable material (U^{235}) is smaller than the critical mass then it is called super critical mass.
8. What is nuclear energy?
   The energy released by a nuclear fission reaction is called nuclear energy or nuclear fission energy.
9. Give any one nuclear fission reaction; mention the factors that impede the nuclear chain reaction.
   $${}_{92}U^{235} + {}_0n^1 \longrightarrow {}_{36}Kr^{92} + {}_{56}Ba^{141} + 3\ {}_0n^1 + \text{energy}$$
   a. The enormous amount of energy released during the nuclear fission is due to the loss in some mass.
   b. During nuclear fission, the sum of the masses of the products formed is slightly less than the sum of masses of target species and bombarding neutron.
   c. The loss in mass gets converted into energy according to Einstein's equation $E = mc^{-2}$
   Where, C = velocity; m = mass loss; E = energy

10. What is mass defect?
The sum of the masses of the products(mp) formed is slightly less than the sum of the masses of the target nuclei (mr) and bombarding neutron. The loss in mass is known as the mass defect, Δm.

11. What is a nuclear reactor?
The arrangement of equipment or device used to carry out the fission reaction under controlled condition is called nuclear reactor

12. What is light water nuclear – power plant?
Light water nuclear power plant is one in which U235 fuel rods are submerged in water. Here the water acts as coolant and moderator. This is called a Light water nuclear power plant.

13. What are moderators? Give some examples.
The substance used to reduce the kinetic energy of the neutrons (1Mev to 0.25 ev) are called moderators. This is done in a small fraction of second. When fast moving neutrons collide with moderator they lose energy and gets slow down.
Example: Heavy water, ordinary water, Graphite, Beryllium.

14. What are fuel rods? Give some examples.
The fissionable materials used in the nuclear reactor are enriched U^{235} and Pu^{239}. The enriched fuel used in the form of rods or strips. It produces heat energy and neutrons which starts nuclear chain reaction.

15. What are control rods? Give some examples.
The reaction rate controlled by movable rods, which made of cadmium or boron if the rods deeply inserted absorb more neutrons reaction, will become very slow.

$$_{43}Cd^{113} + {_0n^1} \longrightarrow {_{43}Cd^{114}} + \gamma \text{ rays.}$$

16. What are coolants? Give some examples.
Coolant: Absorb the heat produced during fission a liquid called coolant is circulated in the reactor core. It enters the base of the reactor and leaves at the top. The heat carried by out going liquid is used to produce steam.
Example: Heavy water, ordinary water, air

17. What is the major role of pressure vessel in the nuclear reactor?
The core and reflector are enclosed in a pressure vessel it provides the entrance and exit passage for coolant. It withstand the pressure up to 200 Kg/cm^2

18. What are fissile nuclides and fertile nuclides?
The fissionable nuclides such as U^{235} and Pu^{239} are readily undergo fission reaction when it is bombarded with neutrons. So they are called fissile nuclides. The nonfissionable nuclides such as U^{238} and Th^{232} which does not undergo fission reaction when it is bombarded with neutrons. So they are called fertile nuclides.

19. Mention any two differences of a nuclear reaction and a chemical reaction.

S. No	nuclear reaction	chemical reaction
1.	It involves a change in composition of the nucleus.	It involves the rearrangement of outer electrons.
2.	New element is produced.	No new element is produced.

20. What are the general components of a nuclear reactor?
The general components of a nuclear reactor are reactor core, fuel rods, control rods, moderators, coolants, pressure vessel, protective shield and turbine.

21. What is breeder reactor?
A breeder reactor is a nuclear reactor that generates new fissile or fissionable material at a greater rate than it consumes such material. It is a device which converts nonfissionable material U^{238} and Th^{232} into fissionable material.

22. Calculate the energy released in the following fission.

$$_{92}U^{235} + {}_0n^1 \longrightarrow {}_{42}Mo^{95} + {}_{57}La^{139} + 2{}_0n^1$$

Given that mass of U^{235} = 235.125

Mo^{95} = 94.945

La^{139} = 138.945

$_0n^1$ = 1.0099

Total mass of products (mp) = $_{42}Mo^{95} + {}_{57}La^{139} + 2{}_0n^1$

= 94.945 + 138.945 + 2(1.0099)

= 235.9098 amu

Total mass of reactants (mr) = $_{92}U^{235} + {}_0n^1$

= 235.125 + 1.0099

= 236.1349 amu

Mass defect = mr – mp

= 236.1349 – 235.9098

= 0.225amu. 1 amu =931.5 Mev

0.2251 amu = 931.5 *0.2251

= 209.7 Mev

23. What is multiplication factor?
The number of neutrons, resulting from a single fission is known as the multiplication factor. When the multiplication factor is less than one, a chain reaction does not take place.

24. Show that 1 amu = 931.5 Mev.
1 amu =
Here, N = Avogadro number = 6.023*1023
Einstein equation, $E = mc^2$

Where, E = Energy; c = velocity of light = 3*108 ms^{-1}

$$E = \frac{1}{6.023 * 1023} *10\text{-}3* (3*108)2\ J$$

$$= 1.49*10 - 10J$$

But, 1Mev = 1.602*10 – 13J

$$E = \frac{1.49 * 10 - 10}{1.602 * 10 - 13}\ \text{Mev} = 931.5\ \text{Mev}$$

25. What is meant by solar energy conversion? How is it done?

The process of direct conversion of solar radiation into useful energy is called solar energy conversion. it can be done by two methods: (i) Thermal conversion, (ii)photo conversion.

26. What is thermal conversion?

It involves the adsorption of thermal energy in the form of IR radiation directly, e.g. solar cooker, solar water heater.

27. Define photo conversion?

It involves the conversion of light energy directly into electrical energy, e.g. solar cell.

28. What is photo galvanic cell or solar cell?

Photo galvanic cell or solar cell is the device, which converts the solar energy into electrical energy.

29. Explain the applications of solar cells.

- Solar cells are used in calculators, electronic watches, street lights, water pumps to run radios and TVs.
- Solar cells are used for lighting purposes
- Solar cells are used to run vehicles

30. What are fuel cells?

Fuel cell is a device which converts chemical energy of the fuel directly into electrical energy without combustion.

Fuel + oxygen $\longrightarrow$ products + electricity.

31. Define fuel cell. Bring out its two applications.

Fuel cell is a device which converts chemical energy of the fuel directly into electrical energy without combustion

Fuel + oxygen $\longrightarrow$ products + electricity

- It is used as auxiliary energy source in space vehicles, submarines, military vehicles.
- The water produced by the cell is valuable source of fresh water by the astronauts

32. What are the reactions involved in oxygen – Hydrogen fuel cell?
- At cathode –electrons absorbed by O_2 and H_2O to produce hydroxide ions

$$O_2 + H_2O + 4e^- \longrightarrow OH^-$$

- At anode-hydrogen molecules oxidized at anode

$$2H_2 + 4OH^- \longrightarrow 4H_2O + 4e^-$$

- Overall cell reaction

$$2H_2 + O_2 \longrightarrow 2H_2O$$

- The emf of the cell is 0.8 to 1 V

33. What are the applications of H_2^- O_2 fuel cell?
- It is used as auxiliary energy source in space vehicles, submarines, military vehicles.
- The water produced by the cell is valuable source of fresh water by the astronauts.

34. What is wind energy? How is it obtained?

The fast moving air is called wind. Energy generated from the force of the wind is called wind energy. The wind have lot of energy because of its high speed. The wind energy is stored in form of electricity through wind mills.

35. What are the drawbacks of wind energy?
- The direction of wind is not predictable
- It affects the television receptor if one lives nearly
- It produces noise.

36. Write short notes on wind energy.

The fast moving air is called wind. Energy generated from the force of the wind is called wind energy. The wind have lot of energy because of its high speed. The wind energy is stored in form of electricity through wind mills.

37. What is a battery? How does it differ from a cell?

It is an arrangement of several electrochemical cells connected in series or parallel, which can be used as a source of direct electric current at constant voltage. Cell is a single arrangement of two electrodes and electrolytic solution capable of yielding electricity during chemical reaction within the cell.

38. What are the important requirements of a battery?
- It should be light and compact for easy transport
- It should have long life.
- The voltage of the battery should not vary appreciably when it is in use.

39. What is a primary battery? Give an example.

Or

What are primary cells?

In these cells the electrode reactions cannot be reversed by passing external electrical energy. The cell reaction occur only once and after use they become dead. Therefore they are not rechargeable, e.g. dry cells and mercury cells.

40. What are secondary cells?

In these cells the electrode reactions can be reversed by passing external electrical energy. The cell reaction occur again and again hence they can be recharged by passing external electric current. They are also called as storage cells, e.g. lead storage cell and Ni – Cd cells?

41. What is a storage cell?

A storage cell is the one which can operate both as an electrolytic cell and as a voltaic cell. When it acts as a voltaic cell, it supplies energy and becomes run down. when it is recharged the cell operates as an electrolytic cell.

42. Write the overall equation for the reaction taking place in an alkaline battery.

Cell reactions

At anode: $Zn\ (s) + 2OH^- \longrightarrow Zn(OH)^2 + 2e^-$

At cathode: $2MnO_2 + H_2O(I) + 2e^- \longrightarrow 2OH^- + Mn_2O_3$

Overall reaction

$Zn\ (s) + 2MnO_2 + H_2O(l) \longrightarrow Zn(OH)_2 + Mn_2O_3$

43. What are the advantages of alkaline battery over dry battery?
   - Zinc does not dissolve readily in alkaline medium.
   - It gives constant voltage when current is drawn from it.
   - The lifetime of alkaline battery is longer than the dry battery because zinc cylinder is not involved in cell reaction, there is no corrosion in zinc cylinder.

44. Write the cell representation of lead storage cell?

Cell representation of lead storage cell

$$Pb\ /\ PbSO_4\ //\ H_2SO_4(aq)\ /PbO_2\ /\ Pb$$

45. Write the charging and discharging reaction of lead accumulator.

Cell reactions

At anode: $Pb\ (s) + SO_4^{2-} \longrightarrow PbSO_4\ (s) + 2e^-$

At cathode: $PbO_2(s) + SO_4^{2-} + 4H^+ + 2e^- \longrightarrow PbSO_4 + 2H_2O$

Overall reaction: Discharging

$$Pb\ (s) + PbO_2(s) + 2H_2SO_4 \longrightarrow PbSO_4 + H_2O + \text{energy}$$

Overall reaction: Charging

$$PbSO_4 + H_2O + \text{energy} \longrightarrow Pb\ (s) + PbO_2(s) + 2H_2SO_4$$

46. What are the applications of lead acid storage cell?
   - It is used to supply current mainly in automobiles such as cars, buses, trucks, etc.
   - It is also used in gas engine ignition, telephone exchanges, hospitals, power stations.

47. Write the cell representation of NICAD battery?

cell representation of NICAD battery

$$Cd\ /\ Cd(OH)_2\ //\ KOH(aq)\ /\ NiO_2\ /\ Ni$$

48. What are the advantages of NICAD battery?
Advantages of NICAD battery
- It is small, light in weight and compact
- The life time of battery is very high as compared to lead storage cell
- It can be packed in a sealed container, but it is more expensive than lead storage cell.

49. How are anodic and cathodic electro active materials made in Ni – Cd battery?

Or

How is NICAD battery constructed?
Anode: Cadmium
Cathode: Paste of NiO_2
Electrolyte: KOH(aq)
Cell representation of NICAD battery

$$Cd / Cd(OH)_2 // KOH(aq) / NiO_2 / Ni$$

Cell produces 1.4 V

50. Describe lithium battery?
a. It consists of a lithium anode and a TiS_2 cathode.
b. A solid electrolyte generally a polymer is packed in between the electrodes.
c. The electrolyte permits the passage of ions but not electrons

Cell reactions

At anode: $Li(s) \longrightarrow Li+ + e-$

At cathode: $TiS_2 + e^- \longrightarrow TiS^{2-}$

Overall reaction

$Li(s) + TiS_2 \longrightarrow Li^+ + TiS^{2-}$

51. Lithium battery is the cell of future, why?
- lithium is light weight metal, only 7 g lithium is required to produce 1 mole of electrons
- The battery can be made in variety of sizes and shapes
- Since all the constituents of battery are solids there is no risk of leakage from the battery
- Lithium cell voltage is high '3V' because Li has the most negative std redn potential. so it generates a higher voltage than all other types of cell.

Answer all questions (16 marks)

1. What is a nuclear reactor? Describe the components of a light – water nuclear power plant with a suitable block diagram.
- The arrangement of equipment or device used to carry out the fission reaction under controlled condition is called nuclear reactor.
- Light water nuclear power plant is one in which U^{235} fuel rods are submerged in water. Here the water acts as coolant and moderator. This is called a Light water nuclear power plant.

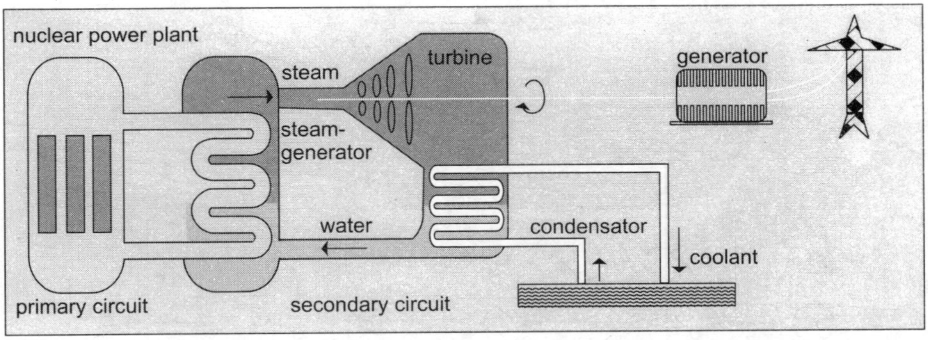

Fig. 15

- *Reactor core:* principle component of nuclear reactor
- *Fuel rods:* fissionable materials enriched U^{235} and Pu^{239}
- *Control rods:* reaction rate controlled by movable rods-deeply inserted absorb more neutrons $_{43}Cd^{113} + _0n^1 \longrightarrow _{43}Cd^{114} + \gamma$ rays.
- *Moderators:* The substance used to reduce the kinetic energy of the neutrons (1Mev to 0.25 ev)
  - o For example, Heavy water, ordinary water, Graphite, Beryllium.
- Coolant: absorb the heat produced during fission
  - o For example, Heavy water, ordinary water, air
- *Pressure vessel:* It provides the entrance and exit passage for coolant. It withstand the pressure up to 200 Kg/cm^2
- *Protective shield:* Covered by thick massive concrete shield to protect from harmful radiations
- *Turbine:* Drives a generator to produce electricity.

2. With a neat sketch explain the functioning of a nuclear reactor.
   - Light water nuclear power plant is one in which U^{235} feel rods are submerged in water. Here the water acts as coolant and moderator (See also Fig. 15).
   - The fission reaction is controlled by inserting or removing the control rods of B_{10} automatically from the spaces Inserted between the fuel rods
   - The heat emitted by U^{235} in the fuel core is absorbed by the coolant
   - Heat is transferred to sea water and then converted into steam.
   - The steam then drives the turbines, generating electricity.

3. Write in brief about chain reaction and nuclear fission reactions.
   - Nuclear fission it is the nuclear reaction in which heavy isotopes are split into lighter nuclei on bombardment by neutrons. Fission reaction of U^{235} is given below
   - The process of propagation of a fission reaction by multiplication in three at each fission is called nuclear chain reaction.
   - Critical mass: Minimum amount of fissionable material required to continue chain reaction.

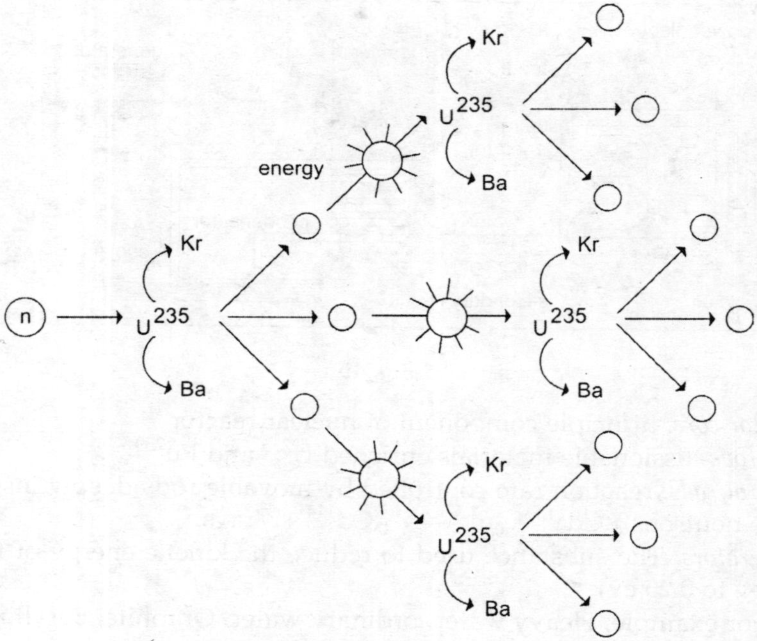

Fig. 16

- Super critical mass: If the mass of fissionable material (U^{235}) is more than the critical mass then it is called super critical mass.
- Sub critical mass: If the mass of fissionable material (U^{235}) is smaller than the critical mass then it is called super critical mass.
- Illustration: ${}_{92}U^{235} + {}_{0}n^{1} \longrightarrow {}_{36}Kr^{92} + {}_{56}Ba^{141} + 3\ {}_{0}n^{1} + \text{energy}$

4. What are breeder reactors? Explain with an example.
   - A breeder reactor is a nuclear reactor that generates new fissile or fissionable material at a greater rate than it consumes such material. It is a device which converts nonfissionable material U^{238} and Th^{232} into fissionable material.
   - The fissionable nuclides such as U^{235} and Pu^{239} are readily undergo fission reaction when it is bombarded with neutrons. so they are called fissile nuclides.
   - The nonfissionable nuclides such as U^{238} and Th^{232} which does not undergo fission reaction when it is bombarded with neutrons. So they are called fertile nuclides.
   - ${}_{92}U^{238} + {}_{0}n^{1} \longrightarrow {}_{92}U^{238} \xrightarrow{-\beta} {}_{93}Np^{239} \xrightarrow{-\beta} {}_{94}Pu^{239}$
5. What are nuclear chain reactions? Explain how to improve the amount of nuclear energy with illustration.
   - The process of propagation of a fission reaction by multiplication in three at each fission is called nuclear chain reaction (see also Fig. 16).
   - Critical mass: Min.amount of fissionable material required to continue chain reaction.

- Super critical mass: If the mass of fissionable material (U^{235}) is more than the critical mass then it is called super critical mass.
- Subcritical mass: If the mass of fissionable material (U^{235}) is smaller than the critical mass then it is called super critical mass.
- Illustration : ${}_{92}U^{235} + {}_{0}n^{1} \longrightarrow {}_{36}Kr^{92} + {}_{56}Ba^{141} + 3\,{}_{0}n^{1}$ + energy

6. State the principle and application of solar batteries.

Ans: Photo galvanic cell or solar cell is the device, which converts the solar energy into electrical energy.

Principle

- Solar cells consists of a p-type semiconductor(Si with B) and n-type semiconductor(Si with P)
- When solar rays fall on the top layer of p-type semiconductor, the electrons from the valence band get promoted to the conduction band and cross the p-n junction into n-type semiconductor.
- Thereby potential difference between two layers is created, which causes flow of electrons.

Application

- Solar cells are used in calculators, electronic watches, street lights, water pumps to run radios and TVs.
- Solar cells are used for lighting purposes
- Solar cells are used to run vehicles.

7. Write a note on fuel cells.

- Fuel cell is a voltaic cell, which convert chemical energy of fuels directly into electricity without combustion.

  Fuel + oxygen $\longrightarrow$ oxidation products + electricity

Construction

- Two porous electrode anode and cathode
- Electrodes-compressed carbon + catalyst (Pt, Pd, Ag)
- Electrolytic solytion – 50% KOH
- Electrodes connected through voltmeter

Working

- At cathode –electrons absorbed by O_2 and H_2O to produce hydroxide ions

  $$O_2 + H_2O + 4e^- \longrightarrow OH^-$$

- At anode-hydrogen molecules oxidized at anode

  $$2H_2 + 4OH^- \longrightarrow 4H_2O + 4e^-$$

- Overall cell reaction

  $$2H_2 + O_2 \longrightarrow 2H_2O$$

- The emf of the cell is 0.8 to 1 V

8. What are fuel cells? Explain the construction and working of fuel cells.

- Fuel cell is a voltaic cell, which convert chemical energy of fuels directly into electricity without combustion.

  Fuel + oxygen $\longrightarrow$ oxidation products + electricity

Construction

a. Two porous electrode anode and cathode
b. Electrodes-compressed carbon + catalyst (Pt, Pd, Ag)
c. Electrolytic solytion–50% KOH
d. Electrodes connected through voltmeter

Working

e. At cathode –electrons absorbed by O_2 and H_2O to produce hydroxide ions

$$O_2 + H_2O + 4e^- \longrightarrow OH^-$$

f. At anode-hydrogen molecules oxidized at anode

$$2H_2 + 4OH^- \; 4H_2O + 4e^-$$

g. Overall cell reaction

$$2H_2 + O_2 \longrightarrow 2H_2O$$

h. The emf of the cell is 0.8 to 1 V.

9. What is primary battery? Give an example.

Primary battery: In these cells the electrode reactions cannot be reversed by passing external electrical energy. The cell reaction occur only once and after use they become dead. Therefore, they are not rechargeable.

For example, dry cells, alkaline batteries and mercury cells.

Alkaline batteries

a. Here the powdered zinc is mixed with KOH and MnO_2 to get a gel
b. A Carbon rod acts as cathode. IT is immersed in KOH
c. The outside cylindrical body is made up of zinc

Cell reactions

At anode: $Zn (s) + 2OH^- \longrightarrow Zn(OH)_2 + 2e^-$

At cathode: $2MnO_2 + H_2O(l) + 2e^- \longrightarrow 2OH^- + Mn_2O_3$

Overall reaction:

$$Zn (s) + 2MnO_2 + H_2O(l) \longrightarrow Zn(OH)^2 + Mn_2O_3$$

Uses: It is used in calculators, watches, etc.

10. Write a note on Ni-Cd battery.

Description

- It consists of a cadmium anode.
- A metal grid containing a paste of NiO_2 acting as a cathode.
- KOH is electrolyte

Cell reactions

At anode: $Cd(s) + 2OH^- \longrightarrow Cd(OH)_2(s) + 2e^-$

At cathode: $NiO_2 + 2H_2O(l) + 2e^- \longrightarrow 2OH^- + Ni(OH)_2 (s) + \text{energy}$

Overall reaction

$$Cd(s) + NiO_2 + 2H_2O(l) \longrightarrow Cd(OH)_2(s) + Ni(OH)_2 (s) + \text{energy}$$

Uses

It is used in calculators. Electronic flash units, transistors and cordless appliances.

11. Write a brief note on lead- acid storage cell.

Description

- It consists of number of voltaic cells connected in series
- Pb is anode and PbO_2 is cathode
- Number of Pb plates and PbO_2 plates are connected in parallel.
- Plates are separated from adjacent ones by insulators like rubber or glass fiber.
- This arrangement is immersed in dil. H_2SO_4

Cell reactions

At anode: $Pb\ (s) + SO_4^{2-} \longrightarrow PbSO_4\ (s) + 2e^-$

At cathode: $PbO_2(s) + SO_4^{2-} + 4H^+ + 2e- \longrightarrow PbSO_4 + 2\ H_2O$

Overall reaction

$$Pb\ (s) + PbO_2(s) + 2H_2SO_4 \longrightarrow PbSO_4 + H_2O + \text{energy}$$

Uses

- It is used to supply current mainly in automobiles such as cars. Buses, trucks, etc.
- It is also used in gas engine ignition, telephone exchanges, hospitals, power stations.

12. Explain the construction and workings of lead acid battery.

Description

- It consists of number of voltaic cells connected in series
- Pb is anode and PbO_2 is cathode
- Number of Pb plates and PbO_2 plates are connected in parallel.
- Plates are separated from adjacent ones by insulators like rubber or glass fiber.
- This arrangement is immersed in dil. H_2SO_4

Cell reactions

At anode: $Pb(s) + SO_4^{2-} \longrightarrow PbSO_4\ (s) + 2e^-$

At cathode: $PbO_2(s) + SO_4^{2-} + 4H^+ + 2e^- \longrightarrow PbSO_4 + 2\ H_2O$

Overall reaction

$$Pb\ (s) + PbO_2(s) + 2H_2SO_4 \longrightarrow PbSO_4 + H_2O + \text{energy}$$

Uses

- It is used to supply current mainly in automobiles such as cars. Buses, trucks, etc.
- It is also used in gas engine ignition, telephone exchanges, hospitals, power stations.

13. Write notes on lithium battery.

Description

- It consists of a lithium anode and a TiS_2 cathode.
- A solid electrolyte generally a polymer is packed in between the electrodes.
- The electrolyte permits the passage of ions but not electrons

Cell reactions

At anode: $Li(s) \longrightarrow Li^+ + e^-$

At cathode: $TiS_2 + e^- \longrightarrow TiS^{2-}$

Overall reaction

$$Li(s) + TiS_2 \longrightarrow Li^+ + TiS^{2-}$$

Other types of secondary lithium batteries

i. Li/MnO_2

ii. Li/V_2O_5

iii. Li/MoO_2

iv. Li/Cr_3O_8

14. What is reversible battery?

In these cells the electrode reactions can be reversed by passing external electrical energy. The cell reaction occur again and again hence they can be recharged by passing external electric current. They are also called as storage cells, e.g. lead storage cell and Ni – Cd cells?

15. Describe the construction and working of lead acid storage battery with reaction occurring during charging and discharging.

Description

- It consists of number of voltaic cells connected in series
- Pb is anode and PbO_2 is cathode
- Number of Pb plates and PbO_2 plates are connected in parallel.
- Plates are separated from adjacent ones by insulators like rubber or glass fiber.
- This arrangement is immersed in dil. H_2SO_4

Cell reactions

At anode: $Pb\ (s) + SO_4^{2-} \longrightarrow PbSO_4\ (s) + 2e^-$

At cathode: $PbO_2(s) + SO_4^{2-} + 4H^+ + 2e^- \longrightarrow PbSO_4 + 2\ H_2O$

Overall reaction

$$Pb\ (s) + PbO_2(s) + 2H_2SO_4 \longrightarrow PbSO_4 + H_2O + \text{energy}$$

16. How is NICAD battery constructed? Explain with cell reaction.

Description

- It consists of a cadmium anode.
- A metal grid containing a paste of NiO_2 acting as a cathode.
- KOH is electrolyte

Cell reactions

At anode: $Cd(s) + 2OH^- \longrightarrow Cd(OH)_2(s) + 2e^-$

At cathode: $NiO_2 + 2H_2O(l) + 2e^- \longrightarrow 2OH^- + Ni(OH)_2\ (s) + \text{energy}$

Overall reaction

$$Cd(s) + NiO_2 + 2H_2O(l) \longrightarrow Cd(OH)_2(s) + Ni(OH)_2\ (s) + \text{energy}$$

17. Give an account of 1. Lead –acid and 2. Ni- Cd batteries.

Lead–acid description:

- It consists of number of voltaic cells connected in series
- Pb is anode and PbO_2 is cathode
- Number of Pb plates and PbO_2 plates are connected in parallel.
- Plates are separated from adjacent ones by insulators like rubber or glass fiber.
- This arrangement is immersed in dil H_2SO_4.

Cell reactions

At anode: $Pb\ (s) + SO_4^{2-} \longrightarrow PbSO_4\ (s) + 2e^-$

At cathode: $PbO_2(s) + SO_4^{2-} + 4H^+ + 2e^- \longrightarrow PbSO_4 + 2\ H_2O$

Overall reaction

$$Pb\ (s) + PbO_2(s) + 2H_2SO_4 \longrightarrow PbSO_4 + H_2O + \text{energy}$$

Ni-Cd batteries description:

- It consists of a cadmium anode.
- A metal grid containing a paste of NiO_2 acting as a cathode.
- KOH is electrolyte

Cell reactions

At anode: $Cd(s) + 2OH^- \longrightarrow Cd(OH)_2(s) + 2e^-$

At cathode: $NiO_2 + 2H_2O(l) + 2e^- \longrightarrow 2OH^- + Ni(OH)_2\ (s) + \text{energy}$

Overall reaction

$$Cd(s) + NiO_2 + 2H_2O(l) \longrightarrow Cd(OH)_2(s) + Ni(OH)_2\ (s) + \text{energy}$$

18. Explain the construction and working of a Lithium Sulphur Battery.

Description

- It consists of a lithium anode and a graphite cathode.
- A solid electrolyte generally β-Alumina is packed in between the electrodes.
- Here sulphur is electron acceptor

Cell reactions

At anode: $Li(s) \longrightarrow Li^+ + e^-$

At cathode: $S + 2e^- \longrightarrow S^{2-}$

Overall reaction

$2Li\ (s) + S \longrightarrow 2Li^+ + S^{2-}$

The sulphide ions, formed, react with elemental sulphur to form the polysulphide ion.

Uses

- It is used in electric cars.

19. What is breeder reactor? Explain the principles of it.

- A breeder reactor is a nuclear reactor that generates new fissile or fissionable material at a greater rate than it consumes such material. It is a device which converts nonfissionable material U^{238} and Th^{232} into fissionable material.
- The fissionable nuclides such as U^{235} and Pu^{239} are readily undergo fission reaction when it is bombarded with neutrons. So they are called fissile nuclides.
- The nonfissionable nuclides such as U^{238} and Th^{232} which does not undergo fission reaction when it is bombarded with neutrons. So they are called fertile nuclides.
- ${}_{92}U^{238} + {}_0n^1 \xrightarrow{-\beta} {}_{92}U^{238} \longrightarrow {}_{93}Np^{239} \longrightarrow {}_{94}Pu^{239}$

20. Write notes on wind energy.

Wind power is the conversion of wind energy into a useful form, such as electricity, using wind turbines. At the end of 2008, worldwide nameplate capacity of wind-powered generators was 121.2 gigawatts (GW). Wind power produces about 1.5% of worldwide electricity use, and is growing rapidly, having doubled in the three years between 2005 and 2008. Large-scale wind farms are typically connected to the local electric power transmission network; smaller turbines are used to provide electricity to isolated locations. Utility companies increasingly buy back surplus electricity produced by small domestic turbines. Wind energy as a power source is attractive as an alternative to fossil fuels, because it is plentiful, renewable, widely distributed, clean, and produces no greenhouse gas emissions; however, the construction of wind farms (as with other forms of power generation) is not universally welcomed due to their visual impact and other effects on the environment.

Wind power is non-dispatchable, meaning that for economic operation all of the available output must be taken when it is available, and other resources, such as hydropower, and standard load management techniques must be used to match supply with demand. The intermittency of wind seldom creates problems when using wind power to supply a low proportion of total demand. Where wind is to be used for a moderate fraction of demand, additional costs for compensation of intermittency are considered to be modest. Studies of a Pan European power grid, show that wind can be used to meet e.g. 70% of load, over a wide area of interconnected grids, and then the costs of electricity delivered into the consuming country are comparable to present day power cost.

22. Write an elaborate note on solar cell and its principle

Definition

- Solar cell is the one, which converts the solar energy into electrical energy.

Principle

- Solar cells consists of a p-type semiconductor(SI with B) and n-type semiconductor(Si with P)

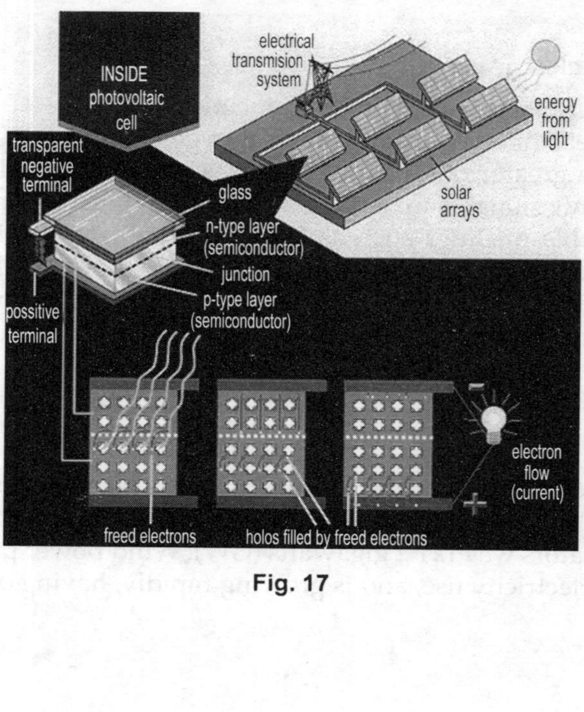

Fig. 17

- When solar rays fall on the top layer of p-type semiconauctor, the electrons from the valence band get promoted to the conduction band and cross the p-n junction into n-type semiconductor.
- Thereby potential difference between two layers is created, which causes flow of electrons

23. What are the difference between nuclear fission and nuclear fusion?

S. No	nuclear fission	nuclear fusion
1.	This process gives chain reaction.	This process does not give chain reaction.
2.	It emits the radioactive rays	It does not emits any kind of the radioactive rays
3.	Spontaneous –ordinary temp	High temp ›106K
4.	Give chain reaction	No chain reaction
5.	Emit neutrons	Emit positrons
6.	controlled	Not controlled

24. Write notes on lithium battery and Lithium Sulphur Battery.

lithium battery Description

- It consists of a lithium anode and a TiS_2 cathode.
- A solid electrolyte generally a polymer is packed in between the electrodes.
- The electrolyte permits the passage of ions but not electrons

Cell reactions

At anode: $Li(s) \longrightarrow Li^+ + e^-$

At cathode: $TiS_2 + e^- \longrightarrow TiS_2^-$

Overall reaction

$$Li(s) + TiS_2 \longrightarrow Li^+ + TiS_2^-$$

Other types of secondary lithium batteries

a. Li/MnO_2

b. $Li/V2O_5$

c. Li/MoO_2

d. Li/Cr_3O_8

Lithium Sulphur Battery Description

- It consists of a lithium anode and a graphite cathode.
- A solid electrolyte generally β-Alumina is packed in between the electrodes.
- Here sulphur is electron acceptor

Cell reactions

At anode: $Li(s) \longrightarrow Li^+ + e^-$

At cathode: $S + 2e^- \longrightarrow S_2^-$

Overall reaction

$$2Li(s) + S \longrightarrow 2Li^+ + S_2^-$$

The sulphide ions, formed, react with elemental sulphur to form the polysulphide ion.

Chapter 5: Engineering Materials

Answer all questions (2 marks)

1. How are refractories classified? Give one example each.
   a. Classified on the basic of chemical propertied of their constituent substances. For example, Acid refractory – Alumina, Silica.
   b. On the basic of refractoriness:
   Low heat duty PCE value -28 -30
   Refractoriness (°C) – 1630 – 1670
   For example, Fire Clay
   c. Based on the oxide content :
   Single oxide refractories – Alumina and Magnesia
2. Mention the characteristics of a good refractory.
   i. Be chemically inert towards corrosive action of glass, slags, metallic liquid, etc.
   ii. Resist the abrading action of flue gases flames, slags, molten metals, etc.
3. What is meant by refractoriness of a refractory?
   It is the property of a material by virtue of which it can withstand high temp. without appreciable softening or deformation under working conditions.
4. What is meant by PCE value of the refractory?
   PCE value of the refractory same is the number of the std pyrcometric cone or reger cone whose apex touches the bottom along with sample cone.
5. Name the stages in the manufacture of refractory.
   Crushing, grinding, screening, mineral dressing, mixing, moulding, drying, fixing or burning.
6. What are the important uses of refractories?
   Used in the construction of lining of furnaces, kilns, crucibles, etc.
7. What is RUL?
   The high temp load bearing characteristics of a refractory material is determined by refractoriness under load test.
8. What are the important properties of high alumina bricks?
   a. Very little tendency to spall
   b. High porosity
   c. Excellent wear resistance.
9. Name any one of the neutral refractories and its chemical composition.
   Chromite – $Cr_2O_3.FeO$
10. How is refractoriness of a refractory m—aterial determined?
    By means of the std PCE test
11. Why is silica bricks expand on heating?
    Because of the irreversible dimensional changes (transformation from one crystalline form into other)

12. Define porosity of a refractory.
    Porosity of a refractory material is given by the ratio of its pores volume to that of its bulk volume.

    $$\text{Porosity, P} = \frac{W - D}{W - A} \times 100$$

13. Mention the objective of PCE test.
    a. To determine the softening temperature of a test refractory.
    b. To classify the fine clay refractories
14. What is meant by dimensional stability?
    It is the resistance of a refractory material to changes in volume due to its prolonged exposure to high temperature.
15. What is meant by thermal spalling? How to avoid it?
    Thermal spalling is breaking, cracking, peeling off or fracturing of a refractory brick or block under high temperature.
    It can be avoided by using high porosity, low coefficient of expansion and good thermal conductivity refractory bricks.
16. What are abrasive?
    Abrasives are the substances used for grinding, cutting, sharpening, polishing, shaping, drilling processes.
17. How are abrasive classified?
    a. Natural abrasive – Silicious and Nonsilicious
    b. Artificial or synthetic abrasives
18. What are soft abrasive?
    The hardness of abrasives are within 1–4 Moh's scale range, then they are known as soft abrasives.
19. What is abrasive power?
    It is the strength of an abrasive to grind away another material. It depends on hardness, toughness and refractoriness.
20. What is hardness of an abrasive? What is their unit?
    It is the capacity of an abrasive to scratch other materials. The unit of hardness is Moh's scale or Vicker's scale.
21. Explain Moh's scale for different abrasives.

Talc	–	1	Fluorite	–	4
Gypsum	–	2	Apatite	–	5
Calcite	–	3	Feldspar	–	6
Corundum	–	9	Quart	–	7
Diamond	–	10	Topaz	–	8

22. Mention some important application of abrasives.
    Abrasives are used as:
    a. Loose powder
    b. Abrasive paper and cloth and
    c. Grinding wheels

23. What is Emery?

It is a fine grained, opaque, black coloured mineral. Its hardness is 8 on Moh's scale.

24. Mention some important characteristics of abrasives.
   a. It should be chemically inert.
   b. It should resist the abrading action.

25. How is carborundum prepared?

$$SiO_2 + 3C \xrightarrow{1500°C} SiC + 2CO$$

26. What is diamond? Mention its types.

It is a pure crystalline carbon. Its hardness is 10 on Moh's scale.

Types of diamond: 1. Black
2. Carbonade

27. How is alundum synthesized?

$$\text{Calcined bauxite + Coke} \xrightarrow[4000°C]{Fe} Al_2O_3$$

28. What is norbide mention its properties?

It is light weight and black coloured material.

Properties: It resist oxidation much better than diamond.

29. What is Garnet? Give its uses.

It is a mixture of trisilicates of alumina, magnesia and ferrous oxide. It is used for making abrasive paper and abrasive cloth.

30. Explain the uses of alundum.
   a. It is used in grinding wheels.
   b. Used in the manufacture of abrasive paper and cloth.

31. Define the terms "Lubricant" and Lubrication".

Lubricant is a substance added in between two moving parts to reduce the friction. Lubrication is a process of reducing friction and wear between two moving parts by adding lubricants in between them.

32. Mention the functions of a lubricant.
   a. It reduces wear, tear and surface deformation.
   b. It reduces wastage of energy.

33. Give two example each for liquid, semisolid and solid lubricants.

Liquid	–	Veg. Oil	–	Palm Oil
Semisolid	–	Greares		
Solid	–	Graphite		

34. Give two examples of synthetic lubricant oil.
   a. Silicones
   b. Polyalycol ethers

35. What are extreme pressure additives? Give examples?

Organic cps – chlorine cpd
Sulphurized oils

36. Define "Oiliness of a lubricant"
The capacity of a lubricants to stick on the surface of the machine parts under heavy load or pressure is called oiliness.
37. Mention some viscosity index improvers?
Polysobutylene, n-hexamol.
38. Define Viscosity index.
The average decrease in viscosity of an oil per degree rise in temperature between 100°F and 210°F
39. What is the significance of pour point of a lubricant?
The pour points indicate the suitability of the lubricants in cold condition and lubricant must have low pour point.
40. Mention the pour-point depressants.
Polyalkyl benzene (para flow)
41. Define the terms "Flash point" and "Fine point"
Flash point is the lowest temperature at which the oil gives enough vapours which ignite for a moment, when a small flame is brought near it.
Fine point is the lowest temperature at which the vapour of the oil burns continuously for atleast 5 seconds, when a small flame is brought near it.
42. What are greases? How are they classified?
Greases are semi solid lubricants obtained by thickening of lubricating oil by the addition of metallic soaps.
They are classified as sodium, calcium, barium, lithium, aluminium and axle (resin) grease.
43. How lithium based grease prepared?
It is prepared by the saponification of vegetable oils or fats with lithiumhyroxide.
44. Mention the different between soda-bare and lime-bare greases. Soda-bare = dropping point of it is very high.
Lime-bare = lesser dropping point than soda-bare grease
45. What are the types of lubrication?
    a. Thick film lubrication.
    b. Boundary lubrication (Thin film)
    c. Extreme pressure lubrication.
46. What are solid lubricants?
The solid lubricants are Graphite and MoS_2.
47. Under what condition extreme pressure lubricant is used?
High load and High speed.
48. What are nano materials?
Nano materials are defined as the materials having atoms/ grains where size ranging from 1–100 nanometers.
49. What are carbon nanotubes?
Carbon nanotubes are large molecules of pure carbon that are long, thin and shaped like tubes about—1–3 nm diameter and hundreds to thousands of nanometers long.

50. How is carbon nanotubes classified?

SWCNT	and	MWCNT
(Single walled)		(Multi walled)

Answer all questions (16 marks)

1. Discuss in detail about any three properties of refractories.
   a. Refractoriness
      - How to measure refractoriness
      - Pyrometric cone equivalent (PCE)

   b. Refractoriness under load (RUL) test
   c. Porosity, $P = \frac{W-D}{W-A} \times 100$
2. Classify refractories into different types and give one example for each type.
   a. According to their chemical properties
      - Acidic refractories – Silica
      - Basic – Magnesite
      - Neutral – Graphite

   b. According to their refractoriness

   Intermediate heat duty — PCC value: 28–30

   Refractoriness 1630 – 1670 **Example,** Fine clay
3. What are the significance of the three characteristics, refractoriness, porosity and thermal conductivity of refractories?

   Refractoriness: To determine the softening temperature of a given refractory material.

   Porosity: A good refractory should have lower porosity

   Thermal spalling: A good refractory should be very good resistance to thermal spalling.
4. What is meant by refractoriness of refractories? How is it determined?

   Refractoriness – Definition

   How to measure refractoriness

   Pyrometric cone equivalent test

   Objectives of PCE

   Measurement of PCE
5. What are refractories? How are they classified? Give example for each type. Explain the term refractoriness and thermal spalling.

   Refractories are inorganic materials that can withstand high temperature without softening or deformation in shape.

   Chemical properties Aluminium

   Classification

   Refractoriness High heat dulty refractory

Refractoriness: It is the ability of a refractory material to withstand at high temperature without softening or deformation under working conditions.

Thermal spalling: Thermal spalling is the breaking, cracking or peeling property of a refractory material under high temperature.

6. How is silicon carbide bricks prepared? Give its important properties and uses.

   SiC particles + binding agent(clay or glue) $\xrightarrow{2000°C}$ SiC

   Properties : i. High strength

   ii. Superior thermal shock resistance

   Uses: Used for partition walls of furnaces

7. What are Garnets and Carborundum? Give the properties of BuC

   Garnets: It is a mixture of trisilicates of alumina, magnesia and ferrous oxide.

   Carborundum: It is a mixture of sand, coke, saw dust and salt.

   BuC: $2B_2O_3 + 7C \xrightarrow{2700°C} BuC + 6CO$

8. Write note on artificial abrasive.

   Carborundum – preparation, prop and uses

   Boron carbide – preparation, prop and uses

   Mundum – preparation, prop and uses

9. Explain the preparation and uses of silicon carbide.

   Preparation: $SiO_2 + 3C \xrightarrow{1500°C} SiC + 2CO$

   Properties: They are mechanically strong and withstand loads in furnaces up to 1650°C

   Uses: Used as heating elements in furnaces in the form of rods and bars.

10. Compare the structure of graphite and MoS_2 and their use in lubrication.

	Graphite	MoS_2
1.	Luminas structure	Sandwitch like structure
2.	Atoms in the layer are bonded by strong covalent bond	Strong covalent bond
3.	Used in lathes, railway track bond	Used in heavy machinery

11. What are greaser? When are they used? Explain the preparation of Li greaser.

    Greaser are semi solid lubricants obtained by thickening of lubricating oil by the addition of metallic soaps.

    They are used where oil is squeezed out due to heavier load or low speed.

    Preparation of Li greaser:

    Saponification of + Lithium hydroxide – Lithium greaser
    veg. oils or fat

12. Write detailed account on the graphite.
    - Graphite consists of multitude of flat layers of hexagonal arrangement of C-atoms.
    - Graphite is very soapy to touch
    - Oil dag is used in internal combustion engines.

13. Indicate the preparation and uses of Li greaser
    - Veg oil + $Li(OH)_2$ – Li greaser
    - It is used at lower temperature. Used for lubricating engines
14. What are solid lubricants? When are they used? Explain the structure of any on solid and lubricant.
    - Graphite and MoS_2
    - They are used where the operating temperature and load is too high.
    - Graphite

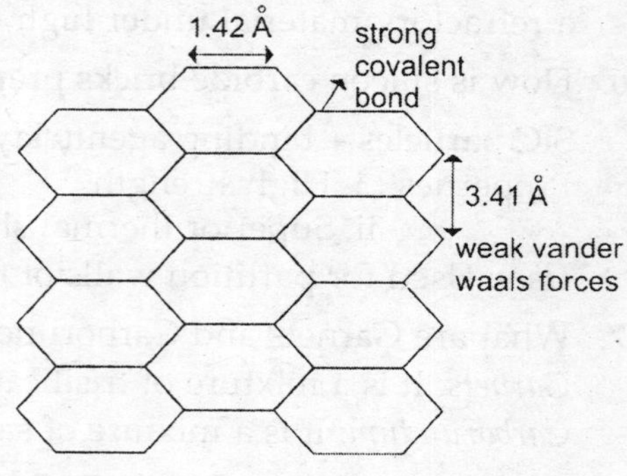

Fig. 18

15. Discuss the importance of viscosity flash point in selecting lubricating oil for a particular use.
    - If the viscosity of the lubricating oil is too high – excessive direction
    - If the viscosity of the lubricating oil is too low – excessive wearer and tear
    - Viscosity index improvers – n-hexanol
    - Pour point depressants – phemols and polyalky, benzene
16. Write notes on flash and fire point of a lubricant.

Flash point – The temp of which the oil gives off enough vapours which ignite for a moment

Fine point – The temp of which the vapours of the oil burns continuously for at least 5 seconds

17. What are carbon nano tubes? Explain their types?
    - Large molecules of pure carbon
    - Shaped like tubes about 1–3 nm diameters
    - 100 to 1000 of nanometers long

Types → SWCNT, → MWCNT

18. Write synthesis, properties and application of carbon nanotubes.

Synthesis of CNT

1. Pyrolysis of HC
2. Laser evaporation
3. Carbon arc method
4. Chemical vapour deposition

Properties of CNT

1. Mechanical
2. Electrical
3. Thermal
4. Vibrational
5. Kinetic properties

Uses or Application of CNT

1. CNT's in storage devices
2. CNT as protective shield
3. Sensors of gases

19. Write notes on the following with examples.
    a. The role of CNT's in catalysis:
        a. $NiO \xrightarrow{CNT} Ni$
        b. $AlCl_3 \xrightarrow{CNT} Al$
        c. $MoO_3 \xrightarrow[MWCNT]{H_2/473°C} MoO_2 + H_2O$
        d. $CdO + H_2S \longrightarrow CdS + H_2O$
    b. CNT's in storage devices
        a. CNT in fuel cells
        b. CNT in lithium battery
20. Explain the structure if CNT

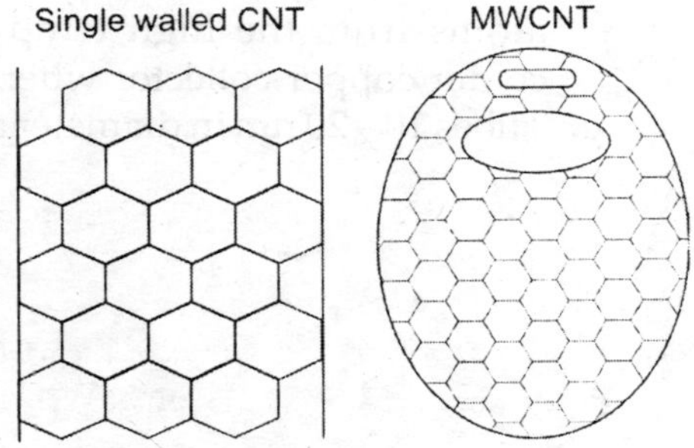

Fig. 19

21. How is pyrolysis carried out?
    Carbon anotubes are synthesis by the pyrolysis of hydrocarbons such as acetylene at about 100°C in the presence of Fe Silica or Fe/Graphite Catalyst under inert conditions.
22. How is chemical vapour deposition carried out?

    Organic gas $CH_4 \xrightarrow[1100°C]{decomposition} CNT$

23. How is carbon arc method carried out?
    - In this method, a potential of 20–25V is applied across the carbon electrodes (graphite electrodes) of 2–50 Mm diameter which is separated by 1Mm at 500 torr pressure of flowing helium.
    - Carbon atoms are ejected from the five electrode and from nanotubes on the negative electrode.
    - The carbon are method can produce single walled nanotubes of diameters 1–5 nm with a length of 1Mm.
24. Discuss about the properties of carbon nanotubes?
    a. Mechanical properties
        - Young's Modulus of CNT's are 10 times greater than that of steel
    b. Electrical properties
        - The electrical properties of CNT's vary between metallic to semi conducting materials.
    c. Thermal conductivity
        - Thermal conductivity of CNT is very high
        - Due to the vibration of covalent bonds which results in very low thermal resistances.

d. Vibrational properties.
   - A1g mode – in and out oscilation
   - $E_{2}g$ mode – squaring of the tube

e. Kinetic properties
   The inner tubes of MWCNT's can slide without any friction within the outer nanotubes leading to an atomically perfect rotational bearing.

25. How is Laser evaporation method carried out?

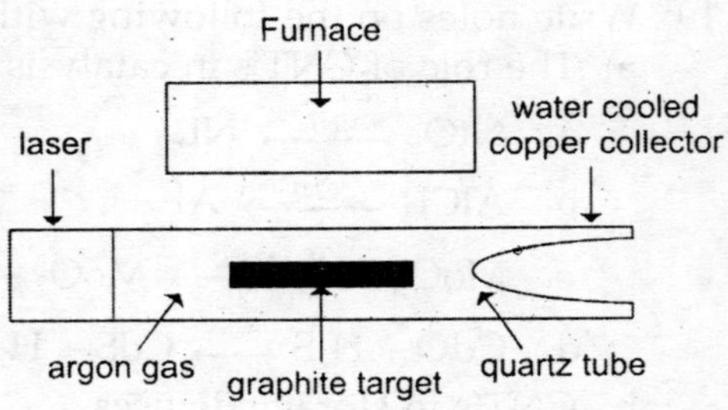

Fig. 20

   a. A Quartz tube contain argon gas and a graphite target is heated to 1200°C
   b. The argon gas carried the carbon atoms from the high temp zone to colder copper collector where it condense into nanotubes.
   c. Tubes 10 – 20 nm in diameter and 100 μm length can be prepared by this method.

Appendix II

Inventions/Discoveries	Inventor/Discoverer	Country	Year
Atom bomb	Rutherford	England	1939
Electrical battery	Volta	Italy	1800
Crystal dynamics	CV Raman	Indian	1928
Discovery of oxygen	JB Priestly	England	1774
Dynamite	Alfred Nobel	Sweden	1862
Laws of gravitation	Newton	England	1687
Radioactivity of uranium	Henry Becquerel	France	1896
Radium	Madam Curie	Poland	1899
Theory of relativity	A Einstein	USA	1905
Telegraph code	Samuel Morse	USA	1832
Theory of evolution	Charles Darwin	England	1859
Telephone	Graham Bell	USA	1876
Wireless(radio)	G Marconi	Italy	1896
X-Ray	WK Roentgen	Germany	1895

INVENTIONS AND DISCOVERIES

Pioneers in Science

- Archimedes—Invention of the famous Archimedean law, discovery of the principles of leaver and specific gravity.
- Avogadro—Italian scientists known for Avogadro's hypothesis.
- Becquerel—Radio activity of uranium.
- Bhabha H J—Research in coxsmic rays and quantum theory.
- Bohr—Electron theory, atomic structure.
- Bunsen—Spectroscope.
- Carothers—Nylon plastics.
- Cavendish—Discovery of hydrogen, rare gases, gases, chemical composition of water.
- Curie, madame—Radium.

- Dalton—Law of partial pressures: Atomic theory, laws of chemical composition: the law of multiple proportions.
- Democritus—Greek philosopher—Atomic theory.
- Einstein—Theory of relativity.
- Euclid—Science of gravity.
- Gay Iussac—Law of gases.
- Herschel, William—Discovered the planet Uranus.
- Hertz—Electrical waves.
- Hoffman—Aniline dye.
- Kelvin, Lord—Dynamic theory.
- Lockyer—Helium gas.
- Maxwell—Electromagnetic theory of light.
- Mendeleve—periodic table.
- Millikin RA—cosmic rays.
- Newton—Laws of motion, laws of gravitation.
- Nobel—Dynamite.
- Otto hahn—Discovery of uranium fission.
- Planck—Quantum theory.
- Thomson JJ—Discovered electrons.
- Urey—Discovery of heavy hydrogen.
- Ramnathan—Molecular scattering of light in fluids.